BY DAN WERB

City of Omens: A Search for the Missing Women of the Borderlands

THE INVISIBLE SIEGE

THE
INVISIBLE SIEGE

THE RISE OF
CORONAVIRUSES
and the
SEARCH FOR A CURE

DAN WERB

CROWN
NEW YORK

Published in the United States by Crown, an imprint of Random House,
a division of Penguin Random House LLC, New York.

CROWN and the Crown colophon are registered trademarks
of Penguin Random House LLC.

Library of Congress Cataloging-in-Publication Data
Names: Werb, Dan, author.
Title: The invisible siege / Dan Werb.
Description: New York: Crown, 2022. | Includes bibliographical
references and index.
Identifiers: LCCN 2021052956 (print) | LCCN 2021052957 (ebook) |
ISBN 9780593239230 (hardcover) | ISBN 9780593239247 (ebook)
Subjects: LCSH: COVID-19 (Disease)—History. | Epidemics—History. |
Virus diseases—History.
Classification: LCC RA644.C67 W47 2022 (print) |
LCC RA644.C67 (ebook) | DDC 614.5/92414—dc23/eng/20211202
LC record available at https://lccn.loc.gov/2021052956
LC ebook record available at https://lccn.loc.gov/2021052957

PRINTED IN THE UNITED STATES OF AMERICA ON ACID-FREE PAPER

crownpublishing.com

ScoutAutomatedPrintCode

First Edition

For those who were not saved

You do all you can to humanize and familiarize the world, and suddenly it becomes more strange than ever.

—Saul Bellow, *The Adventures of Augie March*

CONTENTS

THE INVISIBLE SIEGE

PROLOGUE

March 2020, Seattle, Washington

Nɪᴄᴋ Mᴀʀᴋ ᴡᴀs sᴜʀᴇ ᴛʜᴀᴛ COVID-19 ʜᴀᴅ ʜɪᴛ ᴡᴇᴇᴋs ʙᴇꜰᴏʀᴇ he saw his first official case. In March 2020, the virus had only recently been detected in Seattle, but the sudden outbreaks flared so severely that a first wave must have gone undetected. It was that first wave, he believed, that had entrenched the virus within the city and primed it for its violent surge. As Mark, an ebullient, fast-talking ICU physician, tended to the rising wave of patients, his astonishment at the speed of the virus's path through the hospital quickly shifted to something else: fear.

It began with just one patient, brought into the hospital with a cough and fever, who tested positive for the novel coronavirus. Then, several patients. Within a matter of three weeks, most of the patients at his Seattle-area hospital, the Swedish Medical Center, were there because the virus had colonized their bodies. Mark became shell-shocked. It wasn't so long before that his job was, if not conventional, then at least broadly predictable: as an intensivist, he was trained to care for critically ill patients in the ICU, a role that saw him moving easily through the hospital's hallways, assessing patients as they were brought in and doing his best to keep them alive. It was hard work, but Mark had more energy than most—or perhaps, as he was still in the first couple of decades of his career, his adrenaline was fueling him before some inevitable collapse. Early in the wave, the COVID-infected patients he cared for in the ICU presented with a grab bag of symptoms of varying severity. Some required only brief intensive care and were soon revived after a short course of high-flow oxygen. A small minority, though, was balancing on the knifepoint of death, their immune systems frustratingly resistant to the barrage of treatments (intubation to force oxygen into the lungs, broad-spectrum antivirals, immunosuppressants to keep fevers down) that comprised Mark's usual tools for healing the sick.

Mark prided himself on his ability to cut through the vast emotional gulf that separated patients from their doctors and to use that connection to help those in his care become well again. Each patient was a puzzle, none exactly alike and all requiring his close and focused attention to keep them alive. By late March, though, he was no longer assessing individual patients with particular illnesses and providing them counsel and care; he was dealing with a faceless epidemic. He watched, astonished, as entire wards filled with the COVID infected, and the hospital practically groaned under the weight of the wave, forced to shut down its various functions in favor of one aim: stopping the infected from dying. But as the numbers spiked, keeping people alive became harder. Patients triaged to the ICU presented with severer symptoms, and Mark was plagued with doubt about whether he and his team would actually be able to hold the line as the epidemic unleashed its full force. And that is what had led him to believe that the public health system had missed something. The first flood of patients was so torrential that they could not simply have emerged out of the community all at once. There must have been seeds of the epidemic proliferating before anyone could detect them, the first stones tossed into the water sending ripples, which then caused waves to crash ashore.

While Mark was forced to wage a slow battle of attrition against the virus, SARS-CoV-2 was also, in something of a fifth-column attack, systematically undermining his ability to do his job. Prior to the pandemic, he would move fluidly through the ICU to attend to his patients, doing all he could to make their time in the hospital as free from fear and discomfort as possible. But with the pandemic wave rising, he could no longer perform only a perfunctory washing of hands as he moved from patient to patient. The room itself had become a battle zone, one where the risks of infection had reached the level of life and death. And like a soldier, Mark had to suit up so that he wouldn't be an easy target for the enemy. Preparations began before he entered each patient's room in the ICU. First, he donned a powered air-purifying respirator (or PAPR), a plasticized hood with a built-in fan that sat on his head and shoulders like the chain mail hood on a medieval knight. This contraption created positive pressure to vent particles away from his nose and mouth, the fan connected to a belt around his waist. Then came the synthetic gown placed around his shoulders and tied tightly; plastic booties to cover his shoes lest the virus find purchase there; and two layers of latex gloves. The final accessory was a garden-variety hard hat that hospital technicians had jury-rigged so that Mark and other physicians would be able to see clearly while wearing the PAPR. It was the kind of getup most commonly seen in disaster movies.

Mark knew the personal protective equipment was meant to keep him safe, but he couldn't help feeling anxious every time he was called to a patient's bed and had to suit up yet again. Dressed in his cumbersome gear, he'd wash his hands, suggest treatments he thought (hoped) might inflect the course of their illness, exit the area, wash his hands again, and disrobe. Minutes later, he would have to do it all over again when, say, another infected patient's lungs gave out and they came face-to-face with their imminent death. Again, frustratingly, Mark would have to don his PPE before he could try to help bring them back from the void. There was so much more he could do, he knew, if only he could face the virus without protection.

That wasn't the worst of it. Before the pandemic, Mark was a hands-on physician, knowing full well the power of touch to console patients whose lives were slipping away. Not needing to wear gloves then, he would sit with his patients—many of them elderly—hold their hands, and look them directly in the eye, offering them the gift of skin-on-skin comfort in an otherwise unfamiliar place. All this—the physical contact, his capacity to reveal himself to his patients, to hold their gaze and communicate without words that they were seen, were being cared for, and were, in a way, the object of love—had been made impossible by the threat of a virus that found a home in every body, on surfaces, and in the air itself. "There's a very distant and dehumanizing aspect to it," Mark said. "You're talking to a patient—let's say they're awake—and they can see you're wearing this crazy mask, this crazy outfit. They feel very physically distant from you. Even if you're holding their hand with yours, it's glove on skin as opposed to skin on skin."

In the age of COVID-19, Mark's hospital was no longer just a place of healing. It was, like so many healthcare centers, on the front line in a war against an unthinking and unblinking pathogen with a singular function: to replicate at all costs. That's not to say Mark thought the layers of protective gear and the wards, largely empty except for COVID cases, were overkill. But they were a tangible reminder that he could become the next victim of the novel and enigmatic virus. "You're not accustomed to thinking that way," he admitted. "You're usually thinking, 'I'm trying to save this patient.' Maybe you're thinking, like, 'I'm exhausted' or 'I'm hungry.' You're not thinking, 'If I try to help this person, it could cost me my life.' That's not something I've ever thought about before the pandemic."

Gloved, masked, hooded, and in a flowing sulfur-colored plastic gown, Mark hunched over his patients, day after day and hour after hour, an intubation tube at the ready in one latex-covered hand as he prepared to give them the oxygen they needed. The mouths of his patients were held open, rigid, displaying their hunger for breath, as he passed the smooth, flexible plastic tube across their tongues, Mark trying to concentrate with the whirring of the PAPR's fan buzzing in his ears. He slid the tube farther and farther until he felt the scrape of plastic on the flesh of their tonsils. He'd then gently force it down the esophagus until it rested at the top of the lungs. At that point, mechanical ventilation could begin, with air forced through the tube and into lungs too riddled with viral disease to expand enough to capture oxygen unassisted.

This procedure, tracheal intubation, was one of the basic proficiencies required of Nick Mark in the ICU. Over the span of his career, he had performed it hundreds upon hundreds of times, enough that it had become a series of reflexive actions. But it was only with the threat of COVID literally hanging in the air that the risks of the procedure, which had always been borne solely by the patient, suddenly took on an added dimension. While being intubated, some patients coughed up phlegm and other particulates disturbed by the tube on its rough journey down the human respiratory tract. "Putting a breathing tube in basically sprays viral particles in your face," Mark said. Knowing what might come, he had always taken a mental pause before each procedure—a couple of deep breaths in, a couple of deep breaths out—in an attempt to relax. The ritual worked well when it was only the patient's life on the line. "But I've found that when I was doing that before these coronavirus intubations," he said, "the first time, I was actually more nervous *afterward*, because I was thinking, 'If my PPE fails, I could definitely get this.'" A small tear in his PAPR hood was all it would take.

With each passing day, new victims of the pandemic were added to those already lingering in Mark's ICU, each new patient also a potential vector of the virus. They remained there, many never improving, as the beds filled precipitately with new bodies under his care. Soon, Mark found himself having hushed discussions with colleagues about possibly triaging patients and rationing supplies if the wave of the sick and dying did not abate.

And then the stories began circulating: of shortfalls in PPE, of patients infecting their doctors, of their doctors dying, of entire hospitals being ravaged by COVID-19. It was all too much. Though the virus was basically unknown, and the pandemic in its early stages, it was already obvious to Mark that the path he was on would doom him to become another casualty in the battle against the pandemic.

SARS-CoV-2, THE CORONAVIRUS THAT EMERGED IN 2019 AND THAT, within six months, had spread across the entire globe, was widely described as unprecedented. And while the specific virus that so elegantly burrowed its way into the cells of its human hosts and evaded our species' immune response was a singularly novel threat, it was only the most recent member of an ancient and storied family known as the *Coronaviridae*.

Coronaviruses and humans are not new interlocutors. There are moments in our history—a hundred years ago, perhaps a thousand years ago, or even much longer—when the ancestors of SARS-CoV-2 emerged and caused the same kind of fear, sickness, and death as the pathogen that plagues our time. To call this a comfort would be an overstatement. But there is, in the knowledge that the COVID-19 pandemic is both new and part of a repeating cycle, a way to put boundaries on our anxiety. Indeed, there are ominous precursors throughout human history of the existential dread brought about by the novel coronavirus.

What sets this time apart is that we have come to understand the virus. But just as SARS-CoV-2 is at once new and old, our understanding of it did not start with the first cluster of viral pneumonias in Wuhan. The foundation for our knowledge of SARS-CoV-2 lies in the work of scientists who, decades ago, were driven by a largely abstract fascination with coronaviruses. The study of this viral family, they understood, would yield no plaudits or much recognition; it was simply not that important to humanity. This small cadre of scientists, working in obscurity, nevertheless spent their lives pushing the limits of what could be known about coronaviruses for no other reason than the acquisition of knowledge itself.

Then the family transformed before their eyes into the greatest pathogenic threat of the twenty-first century. As the warning signs flared, the world at large remained blind. And it was then that those

studying the forgotten discipline of coronavirology, a seemingly innocuous subject, became the most consequential scientific minds of
our time. To them we owe the millions of lives that were saved during the COVID-19 pandemic and, if we are lucky, perhaps even the
end of coronavirus pandemics for all time.

Part I

Chapter 1

ONLY OCCASIONAL ECSTASY

February 14, 2003, Hong Kong, People's Republic of China

IT WAS VALENTINE'S DAY, AND MALIK PEIRIS, THE DUTIFUL HUS-band, was out with his wife at their favorite Indian restaurant. It was his first night off in weeks. Peiris, a Sri Lankan–born, Oxford-trained physician and virologist, had lived in Hong Kong since 1995 and was no stranger to emerging epidemics. In 1997, a deadly new strain of avian flu, H5N1, had caused hundreds of illnesses and dozens of deaths in his adopted city, and Peiris had distinguished himself by discovering the mechanism it was using to kill its hosts: a buildup of chemicals called cytokines, which were released by the immune system and ended up causing the body untold damage. But even for someone as well versed in beating back novel viruses as Peiris, the last few months had been bizarre.

Not that Peiris would have admitted as much. His reflexive Old World manners, expressed in a refined British-tinged accent, were accompanied by an innate warmth that put those around him at ease and concealed the intense pressure he put himself under. Peiris carried himself with a compact air, seemingly sewn into a neat suit and tie, with black hair prone to wildness framing a clear-eyed gaze. He betrayed little ego, downplaying with a quiet laugh the severity of the virological phenomena he battled. This amiability had imbued him with a capacity to take setbacks in stride, just as he had done when he was forced to flee Sri Lanka in the face of a growing political insurgency, a move that had eventually led him to his position in Hong Kong. Peiris did not, as a rule, dwell on misfortune. He was as apt to talk about the scientific friendships that forged the path to discovery as he was about how those discoveries might turn the tide in favor of humanity. The problem for Peiris was that in the first few months of 2003, the waves seemed to be rising higher and crashing harder than ever.

Two months earlier, in December 2002, Peiris had been called to two waterparks in Hong Kong that were home to flocks of ducks, swans, and greater flamingos. When he arrived, he encountered utter devastation: dozens of the birds had died, all of them felled by a new pathogen that didn't match any previously detected virus. It was a morbid puzzle, and Peiris was determined to solve it. He immediately canceled his Christmas holidays and spent the ensuing months holed up in his lab at the University of Hong Kong, surrounded by a vast assortment of birds held in an adjacent animal containment facility. Wrestling with the avian sick and dead, he labored to extract samples and placed them inside petri dishes filled with tissue, hoping to get whatever microorganism had spread through the waterparks to replicate. It would take weeks of delicate coaxing, but the pathogen finally bloomed within the confines of his lab, allowing Peiris to isolate it from the surrounding avian tissue and cellular debris. Finally, Peiris gazed upon the specimen under a microscope; immediately, a deep anxiety welled up inside of him. It was a new virus, to be sure, but it looked disturbingly familiar. When he ran further tests, Peiris discovered that, to his horror, the virus was a lethal new strain of H5N1 avian flu, eerily similar to one that had caused dozens of deaths in his adopted city just five years earlier before Peiris spearheaded a citywide effort to contain it.

Peiris passed the next few weeks on pins and needles, waiting, exhausted and alarmed, for the inevitable epidemic to emerge. He had seen avian flu before, but that was cause for fear, not relief. Novel pathogens always started by infecting a cluster of cases, most often in domestic livestock, before the more efficient ones managed to spread across multiple species and into our own. Peiris knew that the waterpark outbreak had all the portents of an emerging killer flu: mass die-offs (the virus was deadly), airborne transmission (it could easily infect multiple individuals), and all of it occurring in environments that humans shared with animals (the virus had ample opportunities to jump from birds to humans). It was, Peiris believed, about to begin again.

And yet, by mid-February, despite his worst fears, the pathogen appeared to burn itself out, by all indications too unfit to replicate efficiently in humans. Peiris, at first cautiously, finally allowed himself to breathe a sigh of relief. The waterpark die-off was not the beginning of the pandemic that he and so many others had feared. It was just another passing virus that had failed to find purchase within the human race.

On Valentine's Day 2003, seated in his favorite Indian restaurant, Peiris looked across the table at his wife and let the moment sink in. After all he had gone through, after all the fear of the last few months, this, finally, was a sliver of happiness.

Then his phone rang.

On the line was the head of the World Health Organization's China mission. He had some troubling news. There was something happening in Guangdong, the Chinese province that neighbored Hong Kong. Hospitals were reporting a cluster of patients with an inexplicable and unusually severe flulike pneumonia. Peiris's heart sank; just when he thought the avian flu outbreak had been self-contained, it appeared that it had indeed jumped the species barrier into humans. He had barely let himself relax, and it was already time to get right back to work.

By the next morning, all the major local newspapers had broken the story of the Guangdong pneumonia outbreak. While neighboring Hong Kong had come under Chinese rule in 1997, the Communist Party still handled its recently reclaimed territory as a pseudo-separate entity. Every day, close to one hundred thousand people crossed the boundary between Guangdong and Hong Kong, passing a formal border as they went. With the pneumonias spreading in hospital rooms and family homes, Peiris knew it was only a matter of time before the outbreaks coalesced into a generalized epidemic. With a reputation burnished from his years of fighting avian flu, he convinced the local authorities to set up enhanced surveillance at the border, which saw agents screening the tens of thousands of crossing visitors for signs of illness. Eventually, they snared two cases: a father and son, both severely ill with pneumonia, who had recently returned from a trip to the mainland for the Lunar New Year festivities. There was a daughter, too, Peiris learned. Like her father and brother, she had become ill during the trip; unlike them, she had perished from her infection before making it back home. Peiris quickly ran samples from both the father and the son through laboratory tests he had designed to detect avian flu strains. When the results came back, he had what he believed was a smoking gun: both were infected with H5N1, the killer flu Peiris had battled in 1997 and that hadn't been seen in humans since.

Peiris assumed that he had found his pathogen. He was wrong. Over the next few weeks, despite hundreds of people becoming ill with the same strange symptoms across Guangdong and Hong Kong, and despite countless tests at the border that separated them, no one else tested positive for H5N1. Remarkably, the father and son's avian flu infection—and the sister's death—was a red herring. With unexplained pneumonias proliferating, it was clear that there was some other unknown pathogen at work. Try as he might, though, Peiris could not identify the virus. He ran every conceivable lab test on samples from sick patients and from those who had succumbed to their illness. Hantavirus, adenoviruses, respiratory syncytial virus, parainfluenza—all were the usual second-line suspects whenever people got sick with respiratory symptoms that couldn't be explained by influenza. All the tests came back negative. Making matters worse, Peiris couldn't get the pathogen to replicate in the routine tissue cultures he grew other respiratory viruses in. That left him working with tiny amounts of virus, far too little to use for the battery of tests he needed to identify the culprit.

Unwilling to concede, Peiris experimented with less common animal tissues to see if he could find some that appealed to the unknown virus. After trying and failing and trying some more, he finally landed on a culture that the virus seemed to enjoy: fetal rhesus monkey kidney cells. Slowly at first, and then with accelerating ferocity, the virus tore through the tissue culture Peiris had gifted it. He finally had a workable specimen of the unknown pathogen.

In the nearly six weeks since he received the phone call from the WHO, Peiris had been working days, nights, and weekends trying to isolate the virus. Every morning and every evening were punctuated with calls organized by the World Health Organization, which had brought together an international network of virologists and epidemiologists to closely monitor the worsening situation. As the pneumonia cluster grew into an outbreak and then metastasized into an epidemic, a sense of dread began to pervade the calls. Since November 2002, when the first cases were recorded among butchers, market vendors, and chefs in Guangdong Province, the strange pneumonia—dubbed severe acute respiratory syndrome, or SARS—had spread far and wide, with no signs of abating. By mid-March 2003, almost two thousand people had fallen ill and hundreds had died across China and the world, the virus having caused outbreaks in Vietnam, Singapore, Taiwan, the Philippines, Canada, and many other countries. No one knew what was causing it, though theories abounded. SARS might have been caused by a retrovirus, which placed it in the same family as HIV, a pathogen that mutates so rapidly and attacks the human immune system so efficiently that, despite billions of dollars and decades of research, it has eluded efforts to create a vaccine. Some thought it was a new kind of herpes, which would have made it a member of one of the most highly contagious viral families—as much as 60 percent of the U.S. population is believed to be infected with herpes—uniquely resilient to the human immune response. Still others, fearing the worst-case scenario, thought it might be a highly transmissible strain of Ebola or Marburg virus, which would place the world on the cusp of a pandemic with a mortality rate upward of 85 percent.

Peiris knew that the path to controlling the SARS epidemic required, more than anything, identifying the true identity of the pathogen. Only then could the tests, treatments, and vaccines needed to stop its spread be developed. With the virus growing happily in rhesus monkey cells, Peiris finally had a shot at isolating the pathogen that had been causing so much sickness and death.

On March 24, 2003, roughly four and a half months after the first SARS cluster emerged, Peiris placed virus-infected tissue culture under an electron microscope and peered through the lens. Under the unblinking beam of subatomic particles, he watched as the cells swayed and buckled under the mysterious pathogen's unrelenting attack. He looked closer, the microscope powerful enough to reveal the shape of the viral intruders. And then came the shock: massive undulating spheres, greeted his gaze. Dotted across their shells were tiny spikes laid out in an orderly lattice, as if crowning the viruses for the victories they had won over their hosts. Peiris had discovered the source of the deadly new epidemic: a virus from an obscure, largely benign family that had never threatened humanity before. Aghast, he had only one question: "Where the devil did this come from?"

February 2003, Vancouver, British Columbia

Every day since the sample arrived, Dr. Bob Brunham had left his lab at the British Columbia Centre for Disease Control and walked two minutes up Ash Street, a verdant, tree-lined avenue flanked by the stark Art Deco tower of Vancouver's city hall to the east and by a low-rise concrete bunker that served as a long-term-care home for seniors to the west. Brunham, a physician and immunologist, had large, guileless eyes set under a mop of curly brown hair. He spoke earnestly, with a soft and tentative voice that suggested that a delicate soul had driven the scientific expertise he had developed over his almost three decades in vaccinology. His destination that day: a provincial cancer agency that was home to the lab of Dr. Marco Marra, Brunham's friend and a member of the city's diminutive class of molecular scientists. Marra's skill set was, in 2003, still something of a scientific novelty: he had mastered the ability to sequence the basic genetic information contained in cells and viruses, their genomes, to better understand how they replicated, so that researchers might exploit their weaknesses. Once Brunham was in Marra's office, the two scientists would excitedly discuss their plan: to map the inner workings of SARS and release their findings to the world.

Brunham had watched from afar as whole cities—Hong Kong, Hanoi, Sydney, and even Toronto—became SARS hot zones and then, as people fled infection, ghost towns. Then the virus landed in Vancouver, his home, and he saw his own colleagues become infected, some of them dying. Brunham could not sit idly by. As a vaccinologist, he knew that the typically decade-long process of creating a viable vaccine couldn't begin in earnest until the virus's genome was mapped. This left the people on the front lines of the epidemic, doctors and nurses, largely on their own and in danger, as there was no proven strategy to care for SARS patients without being exposed to a serious risk of infection. That made Brunham's goal clear: map the virus's genome, scan it to find weaknesses, and develop a vaccine to kill it once and for all.

Brunham and Marra were going to sequence the entire genome of the new, killer virus spreading across the world—and they were going to be the first. After all, as scientists, they were well aware of one of the crueler truths about discovery: nobody cares who solves the riddle second.

WHEN SARS EMERGED, BRUNHAM WAS THE DIRECTOR OF BRITISH Columbia's CDC, a role he had assumed four years earlier, at the age of fifty, after an already storied career as a physician and vaccinologist. To an outsider, his professional path might have made him seem like a dilettante, someone who never quite found what he was looking for. In 1996, he tested chlamydia-prevention programs among sex workers in Kenya. Years later, he tracked a pathogenic fungus that was spreading along the beaches of the Pacific Northwest and killing unlucky porpoises and human beings. And then in 2003, he raced to uncover the genetic code for SARS. Throughout, he had keenly followed the threads of whichever epidemics might need solving at the time, each one leading him back to the singular scientific tool (a method, really) he had been honing for decades: genomics, the science of mapping genes, which made up the building blocks of life. Yet, when SARS emerged shortly after the turn of the twenty-first century, genomics was still something of a novelty.

Brunham's research revealed to him the genetic connections among organisms as diverse as viruses, bacteria, and human beings. It also allowed him to explore the still-mysterious processes by which simple molecules grouped themselves together to become living (and semi-living) creatures. He expressed continual wonderment at being gifted with the chance to travel across these multiple dimensions. "In our world, the information in DNA is something you can't see," he said. "What's very hard to do for many people is to conceptualize what space is like at a molecular level, and how that plays out with all of that molecular motion, and how large these things really are, relative to the size of a water molecule." Brunham was on a journey to bridge these wildly different conceptions of space and magnitude.

Genomics, which uses a combination of complex molecular tests and advanced computer processing to map the structure, function, and evolution of genomes, was Brunham's lodestar. Like archaeology, genomics involves a process of uncovering evidence of life, of separating petrified bones from dust—the difference being that DNA (a double-helix structure that contains the code for all organic life forms) is no fossil. It is a dynamic living record made up of a series of four repeating nucleotides (which are made up of a sugar molecule, a phosphate group, and a unique nitrogen-containing base), and it defines where our species and every other cellular life form on the planet came from and where they might go. Having spent decades deciphering the living record, Brunham had come to believe that life was nothing more than self-sustaining encoded information—and he'd spent his entire career learning to decrypt it.

Revealing the elegant and complex arrangements of molecules was, for many genomic scientists, an end in itself. But Brunham was also a medical doctor who always had one foot firmly planted in the realm of public health. He had come face-to-face with the stark reality that all organisms that carry genetic code—including viruses—have an essential biological mandate: to reproduce as quickly and as efficiently as possible. SARS was evidently realizing that mandate very successfully. In Toronto, it had spread like wildfire: within just one week of the first recorded infection of a healthcare worker, fifty more people had become infected.

While Brunham believed that mapping the SARS genome was the only way to kick-start the path to a SARS vaccine, genomics was in 2003 still marred by a sense of unfulfilled promise. At the time, media coverage of the new scientific discipline was focused exclusively on the Human Genome Project, a "holy grail" pursuit that pitted the privately funded J. Craig Venter, a charismatic and deeply ambitious scientist, against a consortium of publicly funded scientific organizations in a race to map the three billion nucleotide base pairs that make up the entire human genome. Brunham, who had collaborated closely with Venter, was deeply affected by the scientist-entrepreneur's charisma and his drive to unlock the potential for human advancement that lay hidden within our genome. "He had the magic," Brunham said. "When he spoke, he was like a bright light. You wanted to be in his good company," his reverence for Venter clear. "People just surrounded him, including me." One of Venter's chief contributions to genomics was his creation of the "shotgun" method of mapping genomes, which involves exploding DNA into fragments, rapidly analyzing each one simultaneously, and then stitching them all back together. The technique had revolutionized the speed at which scientists could map genomes. Though there was no denying Venter's genius, Brunham had grown disillusioned with his apparent obsession to monetize his discoveries. Doing so, Brunham thought, obscured the Human Genome Project's real potential for helping humanity. So while Brunham was happy that everyone was talking about genomics in 2003, the conversation had become, to his chagrin, focused on how the science could turn a profit rather than about how it could save lives.

Brunham wanted to change that. He was convinced that using genomics to map and then vaccinate against the SARS virus would take the science out of the realm of profit and make it a public good. Before genomics, sequencing DNA (or RNA, a single-helix structure that encodes a small number of viruses) had taken years, because each individual nucleotide base had to be mapped essentially by hand. This made traditional sequencing useless as a means of controlling SARS, which had spread to communities across the world in just three months. But if Brunham and Marra could use genomic science to map the genetic structure of SARS within a matter of weeks, they could trigger a response to the cascading epidemic that could save lives: the virus's genetic code was needed to create a reliable test to diagnose infected people, and it could also be used to track the virus's journey back to its first jump from an animal reservoir. What intrigued Brunham the most, though, was the possibility of using genomics to create a vaccine faster than ever before. Despite the ludicrous odds, he was convinced that he could even create a viable vaccine in the space of a few months or years rather than in the decade or two it had taken scientists to date.

Brunham was well aware that science was, at times, a blood sport and that there were undoubtedly teams across the world also racing to sequence the SARS genome. Some, like those funded by the U.S. Centers for Disease Control and Prevention, were outfitted with hundreds of researchers and with laboratories big enough to fill warehouses. Brunham knew all this and he knew also that the odds of his and Marra's being the first were stacked against them. But he wanted to try anyway.

First, though, they would need a sample of the virus. With SARS spreading rapidly in Toronto, Brunham worked his contacts across Canada to secure a sample, dubbed TOR2, from an unfortunate victim of the epidemic. After a long trip across the country, the sample, a wallet-size Styrofoam box containing a vial of purified viral genetic material, arrived unceremoniously at Marra's lab in Vancouver.

Though it had been harvested using an invasive technique called bronchoalveolar lavage (which uses suction to pull pathogens from lung tissue), its sterile packaging revealed no traces of its origins deep in the respiratory tract of an infected human being. Once the vial was in Marra's hands, its contents was no longer a pathogen with the power to sicken and kill. Suddenly, it was a scientific triumph waiting to happen.

Marco Marra and his team got to work. Theirs was a heavy lift: they had to purify the viral RNA that lay within the TOR2 sample, make a DNA copy of it, and then start a chain reaction known as amplification that would create millions of viral copies, all while trying to outrun the other scientific teams working furiously on their own genetic maps. Though that was difficult enough on its own, TOR2 presented even greater challenges. Marra was an early pioneer of genomic sequencing and was accustomed to being given ample samples when he was cracking genetic codes. But when the TOR2 sample arrived, it left much to be desired. Marra, a tall, lean man with a coiled, confident energy, was taken aback. "We received a very, very tiny aliquot of nucleic acid," he recalled, "like five nanograms, or some astronomically small amount." The sample was so minuscule, in fact, that the liquid in the vial had evaporated, making the vial appear completely empty. Even for a genomicist as accomplished as Marra, five nanograms of genetic material—or five billionths of a gram—was a ludicrously small amount to work with: it weighed about the same as five individual human cells.

The first step for Marra and his team was to figure out whether the virus's genome was made up of DNA, which was relatively durable, or RNA, which was exceedingly fragile. Because the sample had evaporated, Marra's lead technician resuspended it in liquid and then placed some of the mixture onto assays that could detect the presence of either RNA or DNA. The result was a gut punch: it was RNA. Along with the other challenges that were piling up, Marra was now facing a ticking clock as well. RNA is very unstable, meaning that it's liable to break down rapidly in a lab if it comes into contact with other cellular matter or molecules. If any part of the patient were left swimming in the vial among the virus particles, the RNA would disintegrate within a matter of hours.

Marra's team was caught having to carefully balance getting their blind experiments to replicate the virus with needing to maintain enough of the sample's genetic material to map it. Their plan was to use PCR (short for polymerase chain reaction), a technique that allows scientists to rapidly make millions of copies of DNA and RNA fragments that can then be easily analyzed. PCR uses molecules called primers that attach to each end of a segment of DNA or RNA, but for the approach to work, the virus's nucleotide bases (the four molecules that make up every script written into the genome) had to be known. Brunham and Marra, though, had absolutely no idea what the genetic makeup of the TOR2 strain was.

After spending five days trying every technique they could think of to isolate the genetic material from the soup in which it was floating, the team was still at square one. Nothing had worked. "Every time you do something, it takes six hours or overnight," Marra said. "So, you do a thing and then you wait to see it: 'Oh, nothing there.' Do another thing, wait to see it . . ." With each experiment, the odds of successfully mapping the SARS genome dwindled, while the chances rose that the RNA would disintegrate into useless molecules. No matter how impressive anyone's skills might be, there was a limit to how long the team could stave off the inevitable.

Then, on day six, Marra had a brainstorm. If they stitched small bits of a sample of DNA known as a "universal linker" to the ends of the unknown viral genome in the TOR2 sample, they could program the PCR to attach itself to the DNA fragments, which had been fully mapped. If that happened, the PCR would at the very least make copies of the universal linkers. Marra hoped—though there was no guarantee—that the PCR would also copy the unknown parts of the virus's genome sandwiched between the DNA fragments at the same time.

Marra's team delicately added the universal linkers to TOR2, the bundles of atoms securing themselves to the ends of the viral RNA. Then it was time to run the PCR again. By this point, Marra was anxiously aware of how little time they had left before the whole sample degraded into detritus. It was almost certainly their last shot. The team triggered the PCR chain reaction, each cycle doubling the number of copies, the genetic fragments soon numbering in the millions. They then waited for hours, in that liminal, anxious space that lay between undeniable success and abject failure.

When Brunham and Marra checked the results, they almost couldn't believe it. After so many failed starts, the entirety of the TOR2 sample—its complete RNA—had been copied along with the universal linkers. They had done it. Marra was more surprised than anyone. "We got lucky," was all that he would admit. "We guessed right."

Immediately, Brunham and Marra passed the newly replicated copies of the TOR2 strain to a small team of informatics scientists who had been standing by. Their job was to quickly analyze the individual molecules on the RNA as they were being revealed and compare them to the genetic material of known viruses. As Brunham and Marra caught their breath, the lab's computer screens began to light up with that ancient, holy, simple text:

1 atattaggtt tttacctacc caggaaaagc caaccaacct cgatctcttg tagatctgtt

61 ctctaaacga actttaaaat ctgtgtagct gtcgctcggc tgcatgccta gtgcacctac

121 gcagtataaa caataataaa ttttactgtc gttgacaaga aacgagtaac tc-gtccctct

181 tctgcagact gcttacggtt tcgtccgtgt tgcagtcgat catcagcata cctag-gtttc

241 gtccgggtgt gaccgaaagg taagatggag agccttgttc ttggtgtcaa cgaga-aaaca

301 cacgtccaac tcagtttgcc tgtccttcag gttagagacg tgctagtgcg tggcttc-ggg · · ·

The team was shocked: the genome map left no doubt that TOR2 was a coronavirus, even though the only two known human coronaviruses were innocuous pathogens that caused nothing more than mild colds. Weirder still, it most closely resembled, of all things, a bovine coronavirus known mostly for causing fever and dysentery in cattle.

Brunham was floored. He immediately set to work translating what the team had found into information that could help fight the SARS epidemic. The first step was to make the entire sequence completely open, so that any researcher around the world could access it. This wasn't a fait accompli; some members of his and Marra's teams wanted to patent the genome, which would have made it illegal for other scientists or governments to use the sequence without paying for it. Though the SARS epidemic was dominating headlines in April 2003, so, too, was the imminent completion of the Human Genome Project. With J. Craig Venter as a modern-day evangelist for private genomic science, even those scientists most ardently committed to mapping genomes for the public good rather than for profit were having second thoughts about whether theirs was a viable path to scientific discovery. Ultimately, Brunham's team members came up with a compromise: they would patent the SARS sequence but also openly release it, with no strings attached, so that anyone could use it. And so, on April 12, 2003—two days before the announced completion of the Human Genome Project—Brunham and Marra announced that the genome of the SARS pathogen, a novel coronavirus, had been fully mapped.

Marra, Brunham, and their teams had won the race. Less than twenty-four hours later, the U.S. Centers for Disease Control and Prevention announced that it, too, had successfully sequenced SARS. But nobody cares who came in second.

The SARS genome map was the kind of success Brunham lived for. Experimental science was made up, he said, of "pretty repetitive events with only occasional ecstasy." The revelation that a coronavirus was at the heart of the SARS epidemic was one of those profound moments of transcendence. Better yet, he knew that it was the key to combating SARS. With his task completed, he basked in the glow of that rarest of events: the discovery of something entirely new.

But even as Brunham came face-to-face with a new and deadly

pathogen, the revelation that SARS was a coronavirus had implications that went far deeper than the epidemic the world faced in 2003. Coronaviruses were simply not a threat to humans—and they were certainly not supposed to be capable of producing strains that infected eight thousand people across the globe and killed roughly 10 percent of them. In the world of virology, coronaviruses were docile creatures, known for causing common colds and infecting animals. They had, as far as anyone knew, never been anything but a largely innocuous family.

But the sudden emergence of a deadly coronavirus was not as surprising as it seemed. While the SARS virus looked like an impossibility, it was in fact only the most recent progeny of a very old family that had long bedeviled humanity. SARS, for all of its novelty, did not represent a new pathogenic threat. It was only the latest chapter in humanity's long entanglement with the *Coronaviridae*.

Chapter 2

AN EMINENT OFFICIAL WAS
TRAVELING HOME BY BOAT

October 1980, Würzburg, Germany

THE FRANKISH CASTLE WAS IMPOSING, BUT ON THE SMALL SIDE, just big enough for the sixty-odd people who clustered inside its thick stone walls and under its low-slung archways. Everyone there had two things in common: they were all virologists, and they were part of the same small community that represented the totality of human knowledge on coronaviruses. The eclectic group, from labs scattered across the world, had flown to Würzburg (a town with a rather unremarkable history) to swap notes, drink the local Riesling, and share the latest research on their idiosyncratic obsession. There was a certain energy flowing among the group, and while it may not have been the anticipation of a world-shaking discovery, there was nevertheless cause for celebration. The assembled scientists had done it: they had launched the first international conference on coronaviruses despite the fact that the viral family was little more than a scientific oddity. The humble discipline had, in its way, arrived.

"It was a beautiful castle," said Dr. Susan Weiss, a microbiologist whose own forty-plus-year career in coronaviruses was launched during that first conference. Weiss was a slim woman with unruly brown curls and an open, expressive face; when she spoke—on topics as varied as the location of cleavage sites along the coronavirus genome or the beauty of the windswept Southern California coast—it was with a tone that implied that she was sharing deeply held and vital secrets, all delivered with a broad smile. Weiss had an unfailing capacity for wonder, and the trip to Bavaria had proven no exception; Weiss was floored by the opulence and camaraderie she encountered there. Each room featured its own antiques and sumptuous Continental furniture the likes of which she was unaccustomed to. Weiss was a newly minted postdoc at the time, which made her among the most junior scientists at the meeting; in fact, she had found her way into the field only a few months prior by complete happenstance. After finishing her Ph.D. and working in a lab run by the future Nobel Prize winners Mike Bishop and Harold Varmus (the first to discover cancer-causing genes called oncogenes), she had little idea of where she wanted to take her career. All she knew was that she wanted to study viruses. "I literally took a copy of the *Journal of Virology* and thumbed through it, looking for something interesting. And I found this coronavirus." Weiss started to read the short peer-reviewed article and became totally engrossed. What struck her was how much open space there was: no one had yet done any molecular biology work on coronaviruses, which meant that the genetic makeup of an entire viral family was a mystery waiting to be solved.

Coronavirus research was obscure, but this didn't deter her. In those days, virologists who wanted to make a name for themselves studied high-profile pathogenic viruses like influenza, that protean virus that flared with danger signals every few years; or viral hepatitis, three forms of which were discovered throughout the 1970s. "You'd say to people, 'I work on coronaviruses,' and they'd go, 'Huh? What's that?'" Weiss would just laugh. "People would look at you like, 'Who cares?'" It wasn't until SARS, Weiss says, that the world started to pay attention, but even then, most people still didn't know what a coronavirus was. Those decades of obscurity, though, lent the field an enviable openness. "In general, you just had to produce good science. It didn't have to be a therapeutic or a vaccine. It could be just essentially interesting." This meant that Weiss was free to explore whatever she wanted, without the pressure of world-ready scientific deliverables or territorial colleagues. "I thought I'd just do my thing and be happy."

One by one, the scientists in the castle got up to speak about the issues preoccupying them. There were presentations on how best to treat cattle with bovine coronavirus shipping fever; what mouse hepatitis virus (or MHV, a coronavirus widely used in laboratory experiments) could reveal about the root causes of multiple sclerosis; and, perhaps the biggest news of the day, the finding that feline coronavirus, an obscure member of the viral family, could be fatal in some house cats. Other presentations covered coronaviruses in humans, though these were of less interest to the assembled. In 1980, this wasn't surprising: only HCoV-229E and HCoV-OC43, two human coronaviruses discovered in the 1960s, were known to infect people, and both caused only minor colds.

As the scientists clinked glasses, looking out on the Teutonic grandeur of Franconia and the rolling, vine-covered hills that flowed down from the castle, they were, as a field, untroubled. After all, the work they were doing was sure to remain a scientific footnote in the long history of virology, and like most footnotes, it was bound to be ignored.* But many, including Weiss, saw this as a blessing. "I didn't have illusions of, you know, Nobel Prizes or anything," she says. "It wasn't that. It was that the field was ripe for the picking at the time." As they listened attentively and scrawled passing thoughts on notebook pages, the coronavirus experts experienced a studious joy from a realization that they had found their perfect calling: no limelight, little pressure, and lots of room for experimentation away from prying eyes, all for the sake of pure scientific inquiry. The best part, in those halcyon days spent in the Franconian castle, was the near certainty that nothing would ever change that.

* With one exception: A collaborator of Weiss's, Ehud Lavi, had just discovered that mouse coronavirus could induce antigens that attacked brain cells, a fact that had profound implications for research into multiple sclerosis.

CORONAVIRUSES WERE SEEN AS A BACKWATER OF VIROLOGY FOR THE simple reason that their paths, all agreed, would always remain at a distance from our own. Even the fact that a couple of the *Coronaviridae* caused colds in humans wasn't cause for concern; instead, it was seen as proof that the family, and humanity, had found an easy equilibrium. Even so, there were signs that the long period of balance was coming to a close. The arc of every epidemic, no matter its source or the destruction it wreaks, can be traced back to the same cause: the shifting interaction between a pathogen, its host, and the environment within which their relationship plays out. A pathogen may mutate to take advantage of a host in new ways. Hosts (we) might develop new vulnerabilities that allow a previously benign pathogen to colonize and kill. But no matter how perfectly adapted a virus may be to exploit human vulnerabilities, infection can take hold only if a change in its environment causes host and pathogen to collide. The relationships between these three factors—pathogen, host, and environment—are collectively understood as the "epidemic triangle," and they make up a natural law as fixed as gravity. To understand and anticipate epidemics, epidemiologists look to shifts in the points of the triangle that portend a scrambling of the natural static order.

Across the first two decades of the twenty-first century, coronaviruses became the greatest viral threat to our species. Still, they were not new. The cadre of scientists who met in 1980 in the Frankish castle for the first international conference on coronaviruses represented the extent of humanity's knowledge of the family. But the tools of their age weren't powerful enough to probe the history that humanity shared with the viral family, nor could they trace the endless shifts of the epidemic triangle that had brought our species and the *Coronaviridae* hurtling toward each other over and over again through time.

WE LOOK TO THE FACES AND BODIES OF OUR LOVED ONES—OUR friends, our parents, and our children—for evidence of how grueling our journey through time really is. A wrinkle, a gray hair, momentary apprehension when walking up a flight of stairs—if we had no other way to measure time, our bodies would be enough to signal how far we've come and how quickly the path before us is dwindling. And yet, our experience of time isn't uniform, with some days and years lasting longer than others and adding more damage while the young around us grow at speeds that seem close to impossible.

Scientists measure the passage of time in other ways: by the slow appearance of bars in a western blot (a technique for detecting proteins) when nucleotides bind with synthetic antibodies to reveal the presence of viral genetic code; by the speed at which virions (individual viruses) replicate in blood serum, thirstily hijacking human cells to make more of themselves, the assault contained in a petri dish; by the slow unraveling of human DNA under an electron microscope, tight coils delicately touched by proteins that cause them to unfurl, frondlike, and to sway elegantly in a cell's nucleus, their dance directing our bodies in its complex hidden functions. And scientists measure time by the breakdown of their tools: the inevitable obsolescence of their digital sequencers, the accidental fracturing of glass pipettes during routine activity, and the inevitable diminishment of the good hands that once allowed them to work in elegant mastery of their laboratories, which forces them to rely on the next generation to carry out the work.

Viruses age, too, but looking at a single copy of a virus cannot tell you anything about its passage through time. Instead, viral time is measured in the molecular changes introduced during the dance of replication. These mutations, at the level of the virion, are random. But level up to a single human host, with trillions of cells that can be invaded and transformed into staging grounds for the production of billions upon billions of virions, and the speed at which these random errors occur starts to fit a pattern. One magnitude higher, at the level of a virus's spread across the population of an entire species, and patterns of viral mutations become as predictable as clockwork.

While viruses are, by one definition, just simple carbon-based machines, their massive numbers—at least one hundred million different viruses are known to infect vertebrates, invertebrates, lichens, and mushrooms—have led to incredible diversity, not only in their structure and activity, but in the pace at which they evolve over time. On average, a single virion can replicate in just three minutes once it worms its way into an organism's cell. Some highly infectious viruses, like influenza, replicate much faster when they move through new populations, finding weak prey within which they can wreak havoc. Others, like bacteriophages (i.e., viruses that prey on bacteria), coexist in harmony with humans by adding an additional level of protection from deadly invaders. We welcome them into the human virome, that lush garden of viruses living peaceably within our bodies.

In the case of highly pathogenic coronaviruses like SARS and SARS-CoV-2, there is no easy peace to be found. First, individual virions colonize our body's respiratory system, which is made up of about ten billion cells, fertile ground for the viruses to penetrate. Once a virion enters a cell, a countdown to replication begins as the virus releases its genome, which then truculently makes its way through the cell and into the nucleus, where it takes control of the cell's infrastructure and uses it to build copies of itself that then spread anew.

The clockwork errors introduced during viral replication are like those found in every automated, high-volume production technique. Unlike human manufacturing, though, viral errors—which we call mutations—are a key to how viruses find new ways to poke through our bodies' defenses. Through a technique called molecular clock analysis, which calculates the speed at which a virus mutates, scientists can count mutations backward in real time to see the specific year a particular viral strain emerged. It is the viral equivalent of counting the rings in a cross section of a trunk to uncover a tree's age.

All that's needed are a few easily gathered details: a genome map of a virus and that of its closest-known relative and the speed at which the virus mutates. The next step is comparing the number of mutations within DNA or RNA each time that a virus copies itself. As viral genomes unfold themselves, synthesize, and create new copies within human cells, different nucleotide bases are swapped in and out of the long and sinuous structures. Over multiple viral generations, the timing of these random mutations becomes as predictable as the waxing and waning of the moon, allowing scientists to set a clock by them and then to wind it back to see how long it has taken a particular virus, through mutations, to distinguish itself from one of its ancestors. These "molecular clocks" are a surprisingly accurate way of charting when new viral strains emerge, while also giving us the power to tether two wildly contrasting but intersecting narratives: the history of our species and the chance leap of coronaviruses from animal to human hosts.

We now know of seven distinct human coronavirus strains, most of which are benign (though some, of course, are not). Among these are two diminutive viruses that have fully conquered humans: HCoV-OC43, first discovered in 1967 in a cell line made up of organ culture, and HCoV-NL63, first discovered in 2004 in a seven-month-old Dutch infant. The family history of these two coronaviruses—how they evolved and how they became constant companions of the human race—sheds light on what we can expect from the future of the *Coronaviridae*.

After OC43 and NL63 were first identified, copies of the viruses were found in people across the entire globe; by the time they were discovered, they had become, unbeknownst to all, endemic in our species. Nowadays, OC43 and NL63 are believed to be the cause of up to 30 percent of all respiratory infections in humans, most of which manifest as nothing more than a common cold. The combination of the global spread of these viruses and the mildness of the symptoms they cause is a sign that both are very well adapted to their human hosts, meaning that they've found a way to replicate themselves inside our bodies without drawing much attention from our immune system.

But their relative harmlessness is almost certainly a new adaptation. When viruses first make the leap between animal hosts, they are often quite poorly adapted to the new biological systems in which they find themselves, which causes their new hosts to respond vociferously to the new threat. Navigating a new biological system that you're poorly adapted to is a bit like trying to navigate an eighteen-wheel semi-trailer through a low-ceilinged parking garage: you're either going to do it and cause serious damage, or you simply won't be able to do it at all. Yet, over time, evolution can spark trade-offs in pathogens, whereby successful ones mutate to become less deadly to their hosts, while the hosts themselves are able to develop immune responses that systematically weaken the virus so that if it remains in the body it won't replicate enough to become truly dangerous. This broad relationship was quantified in the early 1950s by A. C. Allison, a South African geneticist, during a trip to study the spread of malaria in communities on Mount Kenya. The malaria parasite had become endemic there, but Allison found that the residents were largely unaffected. When he took blood samples from them, he discovered an unusually high occurrence of sickle-cell trait, an inherited and basically harmless form of sickle-cell anemia that caused no major health problems, but that also made the hosts malaria-resistant. This experience sparked Allison's investigation of what became known as the trade-off hypothesis, which posited that viruses and other pathogens would, as epidemics expanded, evolve into less harmful but more transmissible forms. While the trade-off hypothesis doesn't hold universally (the unwavering deadliness of HIV being a rare but important example), it's a useful rule of thumb, especially for viruses that can be transmitted for only a short time by their hosts—like the one- to two-week infectious period that coronaviruses elicit. So, though we know OC43 and NL63 as nothing more than occasional irritants, when they first infected humans, they likely would have resembled their deadlier coronavirus cousins. This, of course, leads to the question of what devastation these two viruses wrought when they first emerged as wholly new creations of nature.

What's clear is that OC43 and NL63 wouldn't have spread to humans without other factors causing their paths to intersect with ours. The epidemic triangle requires just the opposite: there must have been something in the environment, in the virus, or in us that precipitated their leap. The upshot is that while we won't ever know the exact circumstances of the zoonotic jump that saw their spread to humans, we can be sure that their transmission was the result of a dynamic shift in the relationship between the points of the epidemic triangle, which then caused a domino effect, leading to a sudden explosion of new cases.

In the case of OC43, tracing those shifting points of intersection leads us to a curious possibility: that a coronavirus epidemic that killed roughly a million people at the turn of the twentieth century might have been missed entirely.

December 1889, New York City

"SNEEZES FOR ALL. That Fashionable Influenza Will Surely Cross the Sea." On December 13, 1889, this headline—half jest, half warning—was tucked into the top left-hand corner of page 7 of *The New York Sun*. It was written in a period when the passage of time was marked mostly by sunrise and sunset, clocks being still a novelty. It would be almost a century before humanity discovered another type of clock, one whose stuttering ticktock of viral mutations could point to when in time a virus made the zoonotic leap from our animal cousins to us. Sadly for our nineteenth-century forebears, 1889 was the year a molecular clock analysis pinpointed that HCoV-OC43 first collided with man.

"Europe has an epidemic of influenza," *The* (New York) *Evening World* blared, "but do not be frightened. All agree that this latest fad is not dangerous to life—only annoyingly disagreeable to him who has it and excruciatingly funny to him who has it not [*sic*]." And so, in the waning days of 1889, tucked between news articles about twenty-cent Christmas trees and the execution of a Japanese man for a rooming-house murder, the first hints of a pandemic were reported to a U.S. public in terms at once reassuring and dismissive. "The first symptom is a sneeze," the article continued. "Then the nostrils will feel dry. There will be a tightness of the chest. You will cough a dry cough and your throat will be sore. Then you will wonder if life is really worth the bother of living, your spirits will go down, your ambition ooze out at the ends of your fingers and a languor will possess you. You will not care whether school keeps or not for about ten days, then you will brace up, and in another week you will be able to laugh at your neighbor, who is just beginning to sneeze his head off."

The winking irony of the reporting would, in a matter of weeks, abruptly end. Fifteen days after *The Evening World*'s report, on December 28, 1889, Thomas Smith, a twenty-five-year-old from Canton, Massachusetts, became the United States' first recorded "Russian flu" fatality. New York City was particularly hard hit as Americans received glimpses of the full horror of the epidemic racing across the globe. "NEARLY 600 FATAL CASES REPORTED IN PARIS IN ONE DAY," read a headline in *The New York Sun*. "The influenza is spreading and is very fatal," the article stated, accompanied by dispatches from Paris ("Fully one-third of the populace is prostrate"), Vienna ("The hospitals in this city are crowded with patients suffering from influenza"), Lisbon ("Two thousand have the influenza, including the Queen"), and cities across America. In New York, a famed acrobat in a hit play caught "La Grippe"; his manager received a certificate from his doctor stating that he "would probably not live through the night." In Boston, 25,000 people were reported sick with the Russian flu, and authorities said that as many as 10 percent of the city's population had been infected. Ultimately, more than 1,200 New Yorkers and 15,000 people across the United States would die in just the first few months of 1890. The dead numbered over 1 million around the world.

It has long been assumed that the Russian flu was just that: an illness caused by an influenza virus. This was a relatively clear-eyed assumption to make. After all, in the absence of any other major respiratory viruses, influenza was the only suspect. There was also the matter of timing. The Russian flu followed on the heels of successive influenza epidemics in 1729, 1781, and 1830, all of which had emerged at roughly fifty-year intervals, a fact that made it rather predictable that a new influenza would strike in the waning decades of the nineteenth century. (Without microscopes powerful enough to see the structure of viruses—the electron microscope was invented in 1931—scientists had only a dim sense of the pathogens causing these recurring flare-ups.) This influenza hypothesis was bolstered nearly thirty years later, after the Spanish flu pandemic of 1918, which killed 3 percent of the world's population. In the aftermath of the Spanish flu, experts pointed to the Russian flu as its likely viral precursor, a harbinger of doom that humanity failed to heed.

It wasn't until the SARS epidemic, when a pathogenic human coronavirus was first recorded, that scientists started to question whether the twentieth century's obsession with the flu might just have been misplaced. While it was conceivable that the Russian flu was a precursor to the Spanish flu, the two epidemics produced subtly different symptoms. Perhaps most strikingly, the Russian flu had a peculiar telltale symptom: the loss of smell and taste, which was common among those infected with coronaviruses but not those with influenza.

In 2005, with SARS fresh in his mind, the Belgian virologist Marc Van Ranst decided to dig into the origins of HCoV-OC43. While our species has become the natural reservoir for OC43, Van Ranst knew that the virus's closest relative wasn't a human coronavirus. Instead, OC43 was nearly indistinguishable from—or 98 percent identical to—bovine coronavirus, which only infected cattle. This meant that OC43 almost certainly emerged from sustained contact between coronavirus-infected cattle and human beings. By charting the 2 percent of their genomes that was different between the two closely related viruses, and then applying a molecular clock analysis to track how long those changes would have taken to arise, Van Ranst was able to pinpoint the moment when the virus first diverged from its most recent viral ancestor and began to infect our species. He was helped in this effort by Susan Weiss, who had spent the 1980s and '90s identifying which sections of coronavirus genomes were prone to mutation and how key changes made the viruses less deadly. When he ran the molecular clock analysis on OC43, Van Ranst landed on 1890 as the year that it had separated itself from its closest ancestor, bovine coronavirus. Because molecular clock analysis is more reliable the closer the zoonotic leap is to the present, the time period of uncertainty around OC43's emergence was also relatively narrow: between 1859 and 1912, a range of about fifty years. And that's when Van Ranst realized that the evidence pointed to a coronavirus epidemic hiding in plain sight.

The Russian flu was not, in fact, one of the influenza epidemics that regularly appeared every fifty years. Instead, Van Ranst proposed that shifts in the epidemic triangle leading up to 1890 had caused the ideal conditions for a coronavirus to jump from cattle to humans. Beginning in 1850, cattle across the world began to suffer from a disease that caused fever, respiratory tract infections, and dysentery. Because these livestock were transported cheek by jowl, the infection, known as cattle fever, reached pandemic levels. Fevered cattle, driven across wide prairie expanses or shipped for weeks by boat, would drop dead en route; whole herds were felled by the disease. Ultimately, using nineteenth-century technologies, scientists identified the purported agent of disease: *Mycoplasma mycoides*, a deadly and highly transmissible bacteria that spreads through herds via respiratory droplets. But the case was hardly closed. As far back as 2005, researchers pointed out that the symptoms that the bacterium produced were essentially indistinguishable from those of bovine coronavirus infection. And while the bacteria can't infect humans, the fact that OC43 and bovine coronavirus were 98 percent identical meant that the newly emergent CoV could likely—and eventually did—jump from cattle to human beings. (From which source bovine coronavirus first emerged is a mystery, though the virus is closely related to rat, pig, horse, and bat strains.) The novel pathogen just needed the right environment to reach its human hosts. Van Ranst and others suspected that the real cause of the Russian flu had been lurking in cattle herds throughout the late nineteenth century, hidden by the more easily detectable *Mycoplasma mycoides* bacteria.

Beginning in the 1870s, countries across the world began massive culling operations, some of which lasted decades, to eradicate cattle fever. Millions of herds of cattle were slaughtered to stop the spread, which put humans and infected cows in prolonged and intimate contact. As cattlemen in fields and barns held the animals in place, readying them for their death, human and animal breaths comingling as the beasts were dispatched into darkness, both bacteria and virus would have found their way from cow to man—but only one of those pathogens would have been able to flourish in their new hosts. Repeat this scene rolling over two decades and across the world, and the chances of a zoonotic jump move from unlikely to inevitable—but only, of course, if the pathogen had the right tools to survive the journey. OC43, like so many members of its viral family, had those tools. That it succeeded is beyond debate. Billions of copies of the virus circulate freely among us, evidence that it has attained the viral equivalent of immortality: OC43 has become endemic among our species.

SARS PUSHED RESEARCHERS TO PROBE THE ORIGINS OF KNOWN COROnaviruses throughout historical time. It also launched a search for new members—and new threats—emanating from the viral family. And so, while Marc Van Ranst was decoding the byzantine trail of mutations that led OC43 to separate itself from bovine coronavirus, other researchers were stumbling upon evidence of coronaviruses that had also successfully made zoonotic leaps. HCoV-NL63, discovered in 2004 by the Dutch virologist Lia van der Hoek, marked the first time scientists had isolated an endemic human coronavirus since OC43 and 229E (both benign) were discovered in the 1960s. In the wake of SARS, van der Hoek's detection of NL63 revealed a stark reality: our species was more deeply entangled with the *Coronaviridae* than anyone had realized. If the field of coronavirology had been a backwater before SARS, it had become something of a curiosity after the epidemic went global. Still, nobody knew how long the family had been evolving to master our host biology and carve out a home within our bodies.

Van der Hoek wanted an answer, so she cross-referenced the genomic makeup of NL63 with its known relatives and discovered that HCoV-229E was its closest known match. Unlike OC43 and bovine coronavirus, though, NL63 and 229E weren't nearly identical; they were more like distant cousins, sharing only about 65 percent of their total structure. This placed the moment that the two viruses branched off from each other much further back in time. But when? One of van der Hoek's graduate students, Krzysztof Pyrc, set about trying to find an answer. By charting the differences in the genomes of NL63 and 229E, and then watching the speed of each virus's mutations, Pyrc wound the clock back to a wide historical range, 966 C.E. to 1142 C.E., and to a specific year, 1053 C.E., when NL63 emerged out of the zoonotic sea to conquer the human species.

While molecular clock analysis helped with the "when," the question of what kind of carnage NL63 caused when it first emerged was trickier to answer. And unlike OC43, whose emergence in the late nineteenth century lined up neatly with a well-documented global respiratory pandemic, the historical record was far fuzzier in 1053 C.E. There are uncanny signals, though, like an SOS from our distant forebears, that the estimated arrival of NL63 was also a time beset by strange new epidemics. But that's not all: in the eleventh century, one of the most profound transformations in human history was taking place as medieval China planted the seeds of its modern-day future.

1053 C.E., Northern Song Dynasty, China

After a century of chaos and brutality, China was reaching awe-inspiring cultural, economic, and technological heights. The Northern Song dynasty, an imperial era that began in 960 C.E. and ended in 1127 C.E., ushered in a period of political stability and harmonious societal relations after decades of civil war. Hereditary rule by a bloated aristocracy was replaced with a meritocratic central bureaucracy, open to anyone clever and skilled enough to serve. Among this new class of bureaucrats were physician-scientists who rediscovered ancient medical texts with long-forgotten remedies for all manner of ailments and who, in the process, laid the foundation for China's modern healthcare system. With the invention of the movable type printing press around 1040 C.E., these medical texts were copied and recopied, their distribution across the empire nurturing a medical renaissance. It was a time marked by wild scientific innovation and an explosion of knowledge.

Life in the Northern Song was fast-paced and unpredictable. In the capital of Kaifeng, a city of 1.2 million people, it was hard for most to keep up. From a staid administrative center at the beginning of the second millennium, the Northern Song had transformed Kaifeng before its residents' eyes. First, the city's walls, vestiges of medieval warfare, were torn down, giving an airiness to the city that also allowed it to expand in all directions. Then, semipermanent markets moved in, lining the banks of the nearby Yellow River and spilling over into the narrow pedestrian streets, where they blocked horse-and-cart traffic and left the aerosolized tang of wild game, fish, and root vegetables hanging in the air. Kaifeng, built low on the banks of the oft-flooded Yellow River, was cleaved with stone streets and a complex canal system that lent the city a cosmopolitan air, with merchants bringing wares from exotic places (silver, incense, camel, and sheep) along trade routes that penetrated lands more distant from China than ever before. With this revival, the new era also brought about a series of calamities that shocked the empire. The worst, by far, was decades upon decades of near-constant epidemics of strange and incurable respiratory diseases.

It is curious that the estimated range of NL63's emergence (966–1142 C.E.) maps almost perfectly over the dynastic period of the Northern Song (960–1127 C.E.). This may not be coincidence: a dynasty that acted as a political stabilizing force after a century of war, the Northern Song era was marked by radical changes to the third facet of the epidemiologic triangle, the environment. During this period, the population of China boomed from roughly 32 million in 960 C.E. to more than 100 million by the fall of the dynasty in 1127 C.E., almost 30 million more citizens than the Roman Empire at its peak. Kaifeng, the capital, flourished, its million-plus residents densely packed together in close quarters, increasing the risk of rapid transmission of airborne viruses. The conquest and subjugation of territory roughly identical to the modern Chinese state also accelerated the rate of mass migration of people from China's north, the country's traditional heartland, to its south, which has long been a hotbed of emergent infectious diseases, both ancient (bubonic plague) and modern (SARS). And as the Chinese political state stabilized, land and sea routes kept expanding, reaching farther and farther away and bringing contact with new places, new people, and, undoubtedly, new pathogens. In short, it was the perfect environment for an emergent virus to exploit—and a case study in how unchecked epidemics can alter society for decades.

IT WAS A LITTLE OVER TWENTY YEARS INTO THE REIGN OF THE EMPEROR Renzong that the natural laws seemed to become perversely unbalanced. From 1022 C.E. until the mid-1040s, Renzong's rule had been known as the Era of Three Abundances, a succession of nine-year cycles during which the prosperity of the Northern Song expanded, with no limit in sight. But soon there were hints that the emperor's heavenly mandate had been upended. Over a span of twenty years (1041–1060 C.E.), ten major epidemics hit China, a heretofore unheard-of conflagration of disease. But that was just the beginning. The spike in epidemics would ultimately continue over the next eighty-five years, until the demise of the Northern Song dynasty in 1127 C.E., with twenty-eight major epidemics recorded in official court records. Mostly southern in origin, these epidemics spread quickly across the empire, reaching as far as Kaifeng and the emperor's court. (Epidemics have long been seen as among the most inauspicious portents for Chinese rulers, a sign that nature itself was mustering a rebellion against the throne of an illegitimate ruler.)

The cause of the wave of epidemics during the Northern Song will almost certainly remain a mystery for all time. Still, the available clues suggest that it might have been a coronavirus. First, the twenty-year onslaught of epidemics beginning in 1041 C.E. and the molecular clock estimate of NL63's emergence in 1053 C.E. are, in historical and virological time, essentially identical. Beyond the uncanny timing, the radical social transformation (a population that tripled in size over a century, rapid urbanization, and mass migration) created the ideal environment for viral spillover into human hosts. Population growth would have encouraged the taming of forests and jungles, bringing people into contact with wild animals and the pathogens they harbored. The infected hosts would then have returned to increasingly dense cities, ideal sites for seeding outbreaks. The mass movement of people across the vast territory of the Northern Song would also have ensured that localized epidemics spread quickly across the entirety of the empire's vast population.

Emperor Renzong was understandably desperate for a solution to the epidemics that beset his empire. Like all rulers, past and present, his capacity to protect his citizens from nature's wrath was the basic prerequisite for keeping the job. So, he called together China's greatest scientific minds and tasked them with a mission: to track down ancient knowledge that might turn the tide. One philosopher-physician soon hit gold. It turned out that the epidemics the empire was facing fell into a class of diseases known as "cold damage disorders," which had first been recorded in a medical text called *The Treatise on Cold Damage and Miscellaneous Disorders*, dated to around 200 C.E., near the collapse of the Han dynasty.

While there is no easy way to map imperial Chinese medical taxonomy onto modern-day scientific terms, cold damage disorders were, for the most part, bad colds and fevers. By Renzong's time, the study and medical treatment of cold damage disorders had faded into obscurity. This was partly because China had no tradition of inherited, standardized medical knowledge and partly because there just wasn't a need: cold damage disorders were neither widespread nor severe enough to warrant medical attention. This all changed, beginning in the years that NL63 is estimated to have emerged.

"His throat was dry, he had a vexatious thirst ... his body felt hot, and his pulse was small and slightly hurried," wrote Xu Shuwei, a Northern Song physician who compiled ninety case studies of cold damage disorders, this one of a man from Xu's hometown whom he sought to save. Across his medical case reports, Xu described a typical sufferer of cold damage disorders as someone who "sweated spontaneously, his throat was swollen and painful, and he vomited and suffered from diarrhea at the same time...." For case number ninety, Xu wrote that "an eminent official was traveling home by boat ... and the weather was rainy and windy ... and he contracted a disease that resembled a Cold Damage condition." According to Xu, the man's "head was heavy, he sweated spontaneously, and his whole body ached." Xu's assessment, upon seeing the afflicted official, was dire. "This condition cannot be treated," he pronounced. "He died in a very short time."

Xu's ninety case reports on cold damage disorders ran the gamut from jaundice to dysentery to nosebleeds, which aren't necessarily suggestive of any virus in particular. But his reports of cold damage disorders caused by epidemic diseases—which he described as arising from "Warm Winds"—are tantalizingly similar to the set of symptoms caused by human coronaviruses. It is a warm wind that Xu pinpointed as the cause of an epidemic he saw spread after a border conflict with one of China's many hostile neighbors. "When warm wind causes a disorder, there is spontaneous perspiration, the body feels heavy, the patient sleeps a lot, his breathing through the nose makes snoring sounds, and it is also difficult for him to speak." Update the language, and Xu could have written the U.S. Centers for Disease Control's "Symptoms of Coronavirus" list.

Whether NL63 was the cause of the devastating epidemics of the Northern Song or not, the era nevertheless carries lessons on how unchecked pathogens can affect and ultimately cripple a society. One missive from 1086 C.E. paints a foreboding picture of a nation beset by decades of accelerating epidemics. Han Zhihe, a physician, wrote that "thirty years have already passed from 1054 to the present . . . the number of people struck by Cold Damage before the summer solstice has been seven or eight out of ten . . ." Han chalked this issue up to the unwillingness (or inability) of the era's physicians to effectively treat "Yin deficiency," a set of symptoms that included dry cough, sore throat, and fever.

This unending cycle of deadly epidemics evidently took its toll. The celebrated Northern Song–era poet Su Dong-po, widely recognized as among the most brilliant minds in China's long history, was ultimately exiled for his political views, sent to Hainan Island, a barren wilderness off the country's southern coast. After having lost everything (his readership, his political power, his family, and his home), he nevertheless saw exile as a welcome respite. "Here was no meat, no medicine, no houses, no friends, no coal for winter, and no cold spring for summer," he wrote glumly from his island prison. "But there's also one good thing—no epidemics either."

WE CANNOT SAY FOR CERTAIN THAT HCoV-NL63 WAS THE PATHOGEN that drove decades of epidemics across the Northern Song. We can, though, rule some others out. The Black Death, which caused as many as two hundred million deaths, emerged as a pandemic only in the middle of the fourteenth century, roughly two centuries after Emperor Renzong's reign. And though influenza viruses, which cause many of the same symptoms as human coronaviruses, are believed to have been present for at least many thousands of years in China, the first recorded flu pandemic occurred only five hundred years later, in 1580. That pandemic, which started in China, was almost certainly caused by hybrid pig-duck influenza strains that emerged in the fourteenth century when farmers adopted a new farming technique of placing ducks into rice paddies to eat insects, while housing them close to pigs. (Influenza and coronaviruses are only distantly related families, both of them members of the *Orthor-*

navirae kingdom of RNA viruses.) So, while the escalating epidemics of Renzong's reign—which ultimately outlasted the Northern Song dynasty itself—can't be definitively linked to an NL63, they may nevertheless mark one of humanity's first protracted battles with the *Coronaviridae*.

THERE IS A VERSION OF MODERN HUMAN HISTORY IN WHICH THE PANdemic potential of coronaviruses was recognized early. If it had been, the careful work of coaxing lifesaving information out of these bloated, balloon-like virions could have begun with the nineteenth-century Russian flu rather than at the dawn of the twenty-first century during the SARS epidemic. In that other version of history, humanity would as a matter of course have undertaken early and concerted efforts to develop coronavirus vaccines, just as we have for influenzas. That initial failure is the reason Susan Weiss and the small band of coronavirologists who met in 1980 in Bavaria had accepted that their field was and would forever remain a backwater. But that fringe status was based on a fallacy. Coronaviruses, widely considered as curiosities, concealed dark pasts as terrifying human pathogens. And if those secrets had been known, the first international conference on coronaviruses would have been cause not for celebration but for anxiety over what had come before and for terror in the face of the strains yet to arise. But that was not to be.

The sad truth is that innovation in epidemiology, the science of epidemics, is a reductive kind of progress, always facing backward. The epidemiologist's job is to discover the origins of epidemics in order to control them. The harder task is preventing those that have not yet emerged. Each time a novel pathogen creates a foothold in human hosts, we're reminded of just how poor our species is at anticipating the viral threats that we will inevitably face.

The sources of flu contagion have long been obvious. Birds and pigs, familiar animals with which our species coevolved, have since the nineteenth century been known as byways for influenza strains to mix, recombine, and find their way into human beings. But the animal sources of the *Coronaviridae* have proven elusive. Genomic sequencing, DNA manipulation, vaccine development—for all the cutting-edge scientific tools at our disposal, we have had a glaring inability to fully grasp the dangers emanating from this strange and bulbous family and to determine its animal origins. Part of this inability has been due to timing: the Spanish flu, that cataclysmic event, occurred just as our knowledge of viruses and our technology had matured to the point that viruses could be seen. Though it took until the 1930s, the Spanish flu was eventually ID'd correctly (as were two other influenza pandemics, in 1957 and 1968) under microscopes. The absence of pathogenic coronaviruses during this stretch meant that humanity fixed its gaze squarely on the danger it could see: the flu.

Still, our failure to match the threat posed by coronaviruses was part of a broader pattern. For all the scientific successes of the twentieth century, this period was actually marked by a steep and quantifiable decline in the rate of scientific production. By the 1930s, the prevailing assumption was that each new invention would speed the next in a kind of "geometric progression." But throughout the nineteenth and twentieth centuries, the biggest rise in patents per capita in the United States actually took place before 1870, at a time when most of the nation's population was still living in the countryside. In the subsequent century, America's population boom obscured the fact that creating inventions was getting demonstrably harder. While the number of scientific workers in the United States tripled between 1950 and 1993, the country's rate of scientific production remained stable, meaning that it took more people and bigger teams to create the same number of discoveries as before. On average, research productivity dropped 5 percent every year since the 1930s.

Nowhere was this drag on innovation more evident than in the pharmaceutical industry. By the 1950s, morphine, penicillin, aspirin, insulin, and chemotherapy, along with vaccines for polio, measles, smallpox, and tuberculosis, had been developed by or in partnership with private pharmaceutical companies. But the sheer glut of life-saving medicines they produced in the first half of the twentieth

century left them with a choice: either take a chance on more difficult goals, like vaccines for rare or emerging diseases, and risk costly
failures or use their library of intellectual property to make incre-
mental improvements to existing products that were surefire mon-
eymakers. By and large, they chose the safer route. This trend
continued well past the turn of the twenty-first century. For all humanity's awe-inspiring discoveries, our capacity to meet new viral
threats had largely calcified in the face of ruthless market forces.

By 2003, when SARS emerged, the creation of drug therapies
and vaccines for novel pathogens had slowed to barely a trickle. And
so, when, after more than a hundred years of lying dormant, the
Coronaviridae family once again threatened our species, it found us
made vulnerable by hubris, rigidly fixed on the dangers of the past,
and unable to muster our forces to contend with what might lie
ahead. The future, it seemed, had been ceded to our enemies.

THE WEEKS MALIK PEIRIS SPENT ISOLATING THE SARS VIRUS; THE SIX
days Marco Marra and Bob Brunham spent coaxing the viral enigma
to reveal the secrets of its genome; the thread of mutations that
Marc Van Ranst and Krzysztof Pyrc unspooled to pinpoint the moments in our history when deadly new coronaviruses emerged; the
decades Susan Weiss and a small band of coronavirologists spent
working in obscurity, probing deeper and deeper into the mechanics
of how coronaviruses, those microscopic, bulbous machines, overwhelmed their hosts; and the millennia during which the *Coronaviridae* themselves evolved, individual virion by individual virion,
through the random mutations that produced strange new strains
like NL63 and OC43—these are all distinct universes of time, their
magnitudes so different, yet all of them rolling inexorably toward
the moment of our next meeting. SARS surprised the world and
introduced us to the brutal family of viruses ornamented with gaudy
crowns. But it was only the newest face of an ancient enemy. Our
species had encountered deadly coronaviruses before. Each time, it
had paid a price.

But while the family continued its unending replication, recombination, and probing of our species, which culminated in the SARS epidemic, a countervailing force was quietly gaining power. With the aftershocks of each new strain, that force was brought closer and closer to realizing his scientific dream: the subjugation of the *Coronaviridae*. His shadow stretched across the totality of the knowledge that humanity had accrued about this persistent viral family. He was coronavirology's only titan, and his name was Ralph Baric.

Chapter 3

THEY SHOULDN'T EXIST
ON PLANET EARTH

1995, Chapel Hill, North Carolina

UNDER THE VENTED METAL HOOD OF THE BIOSAFETY CABINET, Ralph Baric unscrewed the cap of the plastic rectangular flask and carefully applied a single layer of gelatinous animal tissue. His was a standard biosafety level 2 (BSL-2) lab, and the only viruses present were innocuous enough, but the constant whirring sound of heavy fans venting air and the bright and unflinching fluorescent lights were reminders that it was far from a safe place. Strewn across the tables and glass-paneled workstations were the instruments representing the state of the art in 1990s virology: big, bulky sheets of reactive gels; Sanger sequencing machines, which generated colored bands that revealed a genome's nucleotides one by one; and radioactive autoradiograms producing ghostly X-rays of DNA and RNA. Baric, a former swim champ, had from a young age submerged himself in isolated environments, alone with his thoughts and pushing himself beyond his natural limits. The lab (its equipment, protocols, and routine) was a comfort, a familiar-feeling place where his mind could extend beyond the boundaries of human knowledge and achieve quiet glory.

Baric regarded the flask in his hand, the tissue inside almost transparent. Peering into the microscope, he had an unobstructed view of the individual nuclei of the cells, tiny dark globes within which the stuff of life was created. The cells had been culled from a standard mouse line, the kind that any decent lab could buy off the shelf. The line's rodent progenitor, despite an urgent desire to reproduce, could never have fathomed a second act quite like this: its cells grown and preserved and made immortal to serve the same unavoidable fate, over and over again.

With the biosafety cabinet fans running hard and the monolayer of tissue culture in place, it was finally time. Baric had left a space on the side of the flask next to the mouse cells where he now carefully placed measured droplets of mouse hepatitis virus, or MHV, the coronavirus long used as a safe and fast-replicating laboratory model to study a variety of human diseases. But unlike most scientists, Baric wasn't interested in MHV as a proxy to study something else. He wanted to know about the virus itself, to probe how voracious a killer it really was and how nimble it might be in adapting to whatever shocks it encountered. After adding the virus to the cells, he put the flask on a shelf behind temperature-controlled glass doors. In twenty-four hours, he would check again to see how far the coronavirus had penetrated the cell culture, fulfilling the purpose for which it had evolved since its emergence in the darkness of time.

The next day, Baric examined the MHV virions, which had inched their way relentlessly through the new mouse cells, killing as they went. As it wrought its destruction, the coronavirus created morbidly ornate structures known as syncytia, agglomerations of infected cells fused together by viral particles. Under the microscope, they looked like sunflowers, the desiccated cells resembling petals surrounding a dark mass of their own ejected nuclei. It had a disturbing kind of beauty, but it wasn't what Baric had set up the experiment to create. He wanted the inevitable error and flux, the push and pull of mutations between pathogen and host cells, that he was sure the coronavirus would ignite.

Over days, weeks, and months, at twenty-four-hour intervals, Baric checked the progress of the mouse coronavirus through the cells. As the virus replicated, so did the mouse cell line, and eventually, the inevitable happened: the mouse cells mutated ever so slightly, enough that the virus couldn't conquer them, the genetic variation having created a resistance movement against the would-be conqueror. Remarkably, long after its death the immortalized mouse continued adapting, mutating, and countering new threats, an unyielding force still desperate to live at all costs, in any form. These newly resistant cells, alive amid a landscape of dead ones ravaged by the coronavirus virions, began as a tiny outpost in the plastic flask. But as the weaker cells withered and died, the mutant colony grew, the mouse cells finally getting the upper hand on their microscopic assailants, victory over the virus close at hand.

But just as the mutant cells were saved by chance mutations, random errors in the MHV virions soon inflected the arc of the battle once again. And so, just when the small outpost of mutant cells looked set to dominate the brutal battlefield of the plastic flask, the coronavirus mutated, too, which equipped it with new ways to probe, then attack, and finally kill the mutated mouse cells. Then, pushed to the brink of destruction once again, another small harbor of mutant mouse cells emerged spontaneously, newly resistant to the newest viral variant. Victory once again tilted back toward the cells. And forth. And back. And forth. Over 120 days, Baric would observe the action, cull virus, replace cell lines, and tend to the battlefield in the flask like a constant gardener. "Selection, extinction, and resurrection," Baric said of the experiment, summarizing the eons-old existential struggle of coronaviruses and their living hosts.

In the end, the coevolution in miniature produced a strange new mouse coronavirus variant with uncanny powers. When Baric extracted samples of the experimental virions and tested their hardiness in a range of host environments, he realized that they had learned a stunning trick. The erstwhile mouse coronavirus could now grow in the cells of cats, cows, rats, monkeys, and—troublingly—humans. Baric had created a virus that could infect and replicate, it appeared, in multiple mammalian species. And all it had taken was 120 days. Almost a decade before the SARS virus would begin its brutal assault, his triptych in a petri dish—selection, extinction, and resurrection—was some of the first evidence of the potential pathogenic power of coronaviruses.

A 120-day experiment is, in the human universe, long enough to be memorable. In the universe of coronaviruses, where virions replicate within minutes in countless living hosts drawn from hundreds, if not thousands, of species, it's no time at all. What Baric discovered was that all coronaviruses needed in order to make a comfortable home in a new species was a blink of an eye in viral time. The *Coronaviridae*, pushed to adapt, evolved the tactic needed to meet the moment. Baric had forced the virus into an arms race, and the virus had won. Handily.

If coronaviruses had a track record as a pathogenic threat to humans in 1995, Baric's discovery would have been shocking. But back then, only the cold-causing OC43 and 229E had been discovered. So, Baric's "arms race" experiment ended the way almost all do, with a manuscript published in a peer-reviewed paper that was interesting to others in the field but ignored by the world at large—until, of course, his findings were suddenly wrenched into relevance decades later.

RALPH BARIC WRYLY DESCRIBED HIS CHOICE TO STUDY CORONAVIRUSES as a good example of his farsighted decision-making, "which took about thirty years to come to fruition." In 1982, he had just finished a Ph.D. in microbiology in Raleigh, North Carolina, and was hitting the market as a newly minted junior expert fascinated by viruses. He had gotten there the long way, having literally begun as a dishwasher. When he arrived as an undergraduate on a swimming scholarship, he had taken a job cleaning flasks and other equipment, for which he was paid a couple of dollars an hour. He soon discovered that he loved life in the lab and basically never left, honing his facility with experimentation while blaring the Bee Gees and other disco music as he worked.

Baric no longer had the lithe swimmer's body that had garnered him seven championships and set five collegiate records, but the same quiet intensity that had propelled him across the pool had never left. He still moved at an unhurried pace, letting his arms swing gently by his sides, as if saving his energy for a race. He had long since become barrel-chested, with wisps of thin, graying hair rising lightly from his head, his soft eyes revealing little beyond exhaustion. There was a mannered quietness to Baric's voice, as if he were intent on keeping undisturbed the stillness that enveloped him. Prone to holding his head in his hands and leaning back archly when troubled with complex problems, he would often lapse into prolonged silences mid-conversation, his gravelly baritone fading to an affectless near whisper the closer he came to speaking hard truths. It was evident in those moments of pause that he had transported himself back into the universe of coronaviruses; he would then recall, with uncanny exactitude, specific mutations among tens of thousands of nucleotide bases that caused a particular coronavirus strain to differ from its kin. But there was a self-effacing quality to the codex of knowledge in which he had immersed himself, as if he understood just how ridiculous it was that one person should know that much about so esoteric a subject.

The early eighties were a special time in the field of virology, for one big reason: a new pathogenic virus had emerged, dubbed GRID, for "gay-related immune disease." Laboratories were looking to recruit junior scientists to work on the virus, and Baric had been invited to join the effort. But GRID (later renamed human immunodeficiency virus, or HIV) didn't pique his interest. While it later became humanity's most intractable modern epidemic, in 1982 neither he nor anyone else knew what virus caused GRID or how widespread and devastating it might later become. Baric's gaze was fixed elsewhere, on the largely unknown and ostensibly boring family of benign RNA viruses known as the *Coronaviridae*. In 1982, chasing coronaviruses wasn't going to put him on the map, but he didn't need to be seen. He was captivated by a central mystery at the heart of the family that he had begun to obsess over. It was, in a word, *size*. "They were the largest RNA viruses in nature," he explained, "and theoretically shouldn't exist." These singularly gargantuan viruses presented a paradox, one too intriguing for him to ignore.

RNA, the single helix, is slippery stuff. As Marco Marra well knew when he tried to map the SARS genome, RNA molecules are liable to break down within minutes of being exposed to even the gentlest environmental pressures. Being a single strand of molecules, RNA lacks the inherent structural integrity of DNA, the double helix, and collapses easily in on itself into a pile of molecular junk. What that tenuous architecture means is that every time an RNA virus replicates (which they do constantly), there's a good chance that something will go awry.

Error, though, is a double-edge sword. It can transform an innocuous mouse coronavirus into a multispecies pathogen that easily tears through hamster, rat, cat, cow, monkey, and human tissue. But it can also stop viral replication dead in its tracks. Enough error—which in the genomic universe means that incorrect molecules are swapped into a viral genome as it is being copied—and RNA will end up collapsing in a heap. This phenomenon is called error catastrophe: with the accumulation of enough mutations in a replicating genome, the nascent virus's structure becomes too shaky until it can't hold together and its genetic code is unable to perpetuate itself.

RNA viruses have dealt with the problem of error catastrophe by the most efficient means possible: downsizing. If you aren't able to limit how frequently errors happen during replication, you can at least reduce the overall number of places where they might occur. It's the same basic calculus that explains why it's easier to make a peanut butter sandwich than, say, *Poularde de Bresse truffée en vessie* (chicken with truffles in a pig's bladder); there are just fewer ingredients to mess up. Unable to reduce their rate of mutation, RNA viruses have instead shed as much of their genomes as possible. While a DNA virus like herpes simplex can have a genome 375 kilobases long (meaning that it contains more than 375,000 base pairs), the average RNA virus genome is only 9 kilobases long, with 9,000 individual bases. This lean approach has allowed RNA viruses to flourish despite their predilection for mutation.

Because of this inherent fragility, scientists calculated that RNA viruses wouldn't be able to support genomes larger than 18 kilobases in length. And yet, the genomes of coronaviruses were much, much larger, sometimes stretching to 30 kilobases. Baric was intrigued. "They shouldn't exist on planet Earth," he thought. And yet they did. It was a mystery big enough for him to want to go giant hunting.

It took Baric over three decades to unlock the secret to why coronaviruses had such impossibly large genomes. Like all RNA viruses, the *Coronaviridae* should have encountered physical limitations on the size of their fragile genomes, but the family had somehow broken through those limits as if they didn't exist. As technology advanced, Baric and his colleague Mark Denison, a pediatric immunologist with kind, sloping eyes and an avuncular tone (and most important, a fellow coronavirus obsessive), made it their mission to find out why. The duo spent years scouring the entire genomic structure of coronaviruses, all 30,000 bases, to figure out what each viral component did. It was grueling work.

Coronaviruses, Baric and Denison knew, were made up of two basic sections: structural proteins that held the virus together and a larger set of nonstructural proteins with a variety of functions that helped the viruses replicate—though how they did so remained obscure or unknown. Together, these sections encased the viral RNA, which would be inserted into a host cell and serve as a set of instructions that forced the cell's nucleus to produce more copies of the virus. The structural proteins included a spike, an envelope, and a membrane, all of which made up the oily shell that encased the viral genome, as well as a kind of cage, known as a nucleocapsid, that lay inside the envelope and coiled itself tightly around the viral RNA—the closest thing viruses have to an immortal soul—to protect it from injury.

The sixteen nonstructural proteins were where the mystery lay. Some of these proteins helped the virus replicate once it entered a host cell, while others slowed down an invaded cell's ability to mount an antiviral attack. One protein, known as nsp3 (for "nonstructural protein 3"), removed the tags that a host's immune system placed on viral proteins, thus allowing coronaviruses to evade their host's best efforts at eradicating them. Other nonstructural proteins added camouflage, made bubbles to house viral components, moved molecules around a cell, or cut up and freed viral proteins that were critical to moving replication along. These myriad and complex tasks, carried out by mere bundles of atoms, hinted at the root desire for life that lay at the heart of every organism. But none explained how coronaviruses were able to buck the laws of nature and grow to their enormous size. After searching for years, Baric and Denison finally learned the secret.

The answer, which they discovered in 2007, twenty-five years after Baric started studying coronaviruses, lay in a nonstructural protein dubbed nsp14-ExoN, which was produced by viral RNA during replication. Being "nonstructural," the nsp14-ExoN protein didn't hold the virus together. Instead, it did something that no other RNA virus was capable of: it acted as a proofreader, combing over newly copied strands of RNA, removing random errors in the base pair sequence as it went, and replacing them with the correct molecule, like a worker doing quality control at a manufacturing plant. By doing this, nsp14-ExoN was able to keep the number of mutations to a minimum, which slowed the overall evolution of coronaviruses but allowed them to stretch the size of their genomes.

It had long been known that every DNA-based life form had proofreading proteins that maintained the structural integrity of their genomes and allowed them to grow in complexity. (Human DNA, for example, contains around three billion base pairs.) But it wasn't until Baric and Denison discovered the function of the nsp14-ExoN protein that an RNA virus was shown to do it, too. All RNA viruses, Baric once wrote, "must achieve a balance between the capacity for adaptation . . . [and] the need to maintain an intact and replication competent genome." What the discovery of the coronavirus proofreading mechanism showed was how uniquely static the *Coronaviridae* were compared with the forty-six other known RNA virus families, the rest of which were in a constant, unending state of mutation.

As Baric looked deeper into the viral family, he realized that the viruses' size was only one of their many mysteries. "Nobody really knew how they replicated, how they expressed their genes, what basic mechanism allowed these viruses to exist in nature. All that was unknown." Coronaviruses just weren't that interesting—they were too innocuous—for a scientist who was out to change the world. But Baric wasn't that scientist. Where others saw a career death trap, he saw a completely underappreciated and understudied virus family. Coronaviruses were a limitless ocean of knowledge, uncharted and unsullied, and all Baric wanted to do was dive in.

BARIC'S INTRODUCTION TO CORONAVIRUSES BEGAN WITH A POSTDOC-toral fellowship at the University of Southern California in Los Angeles. He would eventually spend four years there (1982–1986) studying the mechanics of how the family replicated. His two mentors had gravitated toward MHV, the mouse coronavirus, as a useful but temporary tool in careers that mostly spanned other fields. One mentor was an RNA tumor virologist who used coronaviruses as proxies to study his main passion, cancer-causing RNA viruses like human papillomavirus. The other had found endless uses for MHV in studying the fundamentals of how viruses caused human diseases as varied as hepatitis, Parkinson's, and multiple sclerosis, which allowed scientists a straightforward way to develop treatments. During his time in L.A., Baric was trained to manipulate MHV's various mechanisms to see how "knocking out" different genes—essentially, forcefully removing parts of the virus's genome—affected Parkinson's and multiple sclerosis progression in mice. By the time he started a faculty position at the University of North Carolina in March 1986, he had become an expert at manipulating viral genetic material.

The years in Los Angeles also cemented Baric's lifelong obsession with coronaviruses. He would ultimately become their most ardent chronicler, the hundreds of peer-reviewed scientific studies bearing his imprimatur reading like an extended family history. The studies are akin to viral biographies, like the Roman historian Suetonius's *Twelve Caesars*, laying bare the intrigue, deadliness, and complexity that drive these ancient and enigmatic antagonists.

Baric's work with MHV was just beginning. It wasn't enough for him that he had created a highly destructive version of the mouse coronavirus simply by putting obstacles in its path in the form of mutation-prone mouse cells. He now wanted to push the virus to its absolute limit to witness how far its impressive capacity for adaptation could go. In the 1990s, there seemed no reason not to: the field of coronavirology was beyond obscure, the only human coronaviruses caused nothing more than colds, and Baric was a young scientist whose mind was overflowing with ideas.

Baric wanted to create, in his words, "experimental evolution," the transformation of the virus under conditions closely mimicking selection pressure, that confluence of ever-shifting natural forces that rewards useful mutations and punishes futile ones. The point was to see how quickly the virus could adapt if mice, its natural host reservoir, went extinct. He was, in essence, creating his own epidemic triangle in the sanctity of his lab.

Baric went back to the biosafety cabinet, flask in hand, and again applied MHV (which can produce up to ten million individual virions in a single cell) to a thin, translucent layer of mouse cells. He then tried to grow the same virus in a separate flask containing hamster cells but found, unsurprisingly, that it couldn't grow at all. This baseline result confirmed that a virus designed for one species couldn't be expected to jump to another without a little nudge. The next move was obvious. Baric didn't just nudge the virus; he shoved it—hard.

Baric and his team created "cocultures" by mixing a half-and-half swirl of mouse and hamster cells in the same container. Once again, he introduced MHV virions into the mix. Twenty-four hours later, he harvested the progeny virus and infected a new swirl of mouse and hamster cells, but this time he increased the number of hamster cells and decreased the number of mouse cells. A day later, and every day after that for weeks on end, he harvested the virus and again added it to a container with fewer mouse cells and more hamster cells, to mimic a scenario in which one species in a certain area is displaced by another, more dominant species. When that's the case, a virus adapted to the declining species will have to quickly find a new host or disappear. The basic question for Baric was whether coronaviruses would end up disappearing in the face of an extreme environmental shift or whether they would find a way to outrun their own extinction.

After four months of harvesting the virus and adding more hamster and fewer mouse cells, Baric ended up with a strain of mouse coronavirus that could produce one hundred million virions in a single hamster cell—ten times more than its off-the-shelf ancestor was able to produce in the cells of mice, its natural host. It was a sobering outcome, one made more shocking by the speed with which the virus found a way to thrive in a hostile new environment.

Of course, ours is not a world where hamsters are hell-bent on driving mice to extinction. We do, however, live in a world where deforestation, exotic animal hunting, and climate change have caused humans to encroach on pristine natural environments, a situation that has slowly eradicated the species that live there as our inexorable march toward progress has continued apace. In the peer-reviewed paper reporting his findings, Baric noted that the experiment proved "that a relatively few amino acid alterations are needed to alter species specificity in coronaviruses." That meant that coronaviruses weren't just able to jump; they were at the precipice, seconds away from leaping. Baric had, in his quiet way, sounded an alarm. Though his experiment was carried out in 1996, six years before SARS would make its fateful embrace of our species, its results were prescient. "Given the nearly unprecedented transformation in the world's ecology, our findings suggest that RNA viruses . . . may adapt rapidly by episodic evolution," Baric concluded. Coronaviruses, he realized, were adaptive enough to weather environmental chaos and thrive within whichever species ended up dominating the landscape. Us.

The next step for Baric was to find out how MHV had outgunned its fate across the four-month-long marathon to which he had subjected it. He had just conducted multiple experiments proving that, when push came to shove, a formerly benign mouse coronavirus could turn into a pathogen that thrived across multiple species. There had to be some trick that coronaviruses used to parachute safely into new environments.

Baric stood over his lab's bulky Sanger sequencer, watching as sheets of gel reacting to coronavirus genomes spelled out the ancient four-letter script. In the 1990s, there was only one way to map genomes, so Baric watched patiently as thin black bars appeared across four columns in the gel, each bar corresponding to one of four nucleotides, the nucleotide chains making up coronavirus RNA. He transcribed their position by hand, bar by bar, nucleotide by nucleotide, the script numbering in the hundreds, then the thousands, and then the tens of thousands, until the entirety of the genomes was complete. It was long and arduous work, but Baric didn't mind. With a panel of fully sequenced coronavirus RNAs, he could scan their vast genomes for the secret he was chasing: the location of the skeleton key that allowed the *Coronaviridae* to unlock entry into strange and varied animals.

THE FIRST DESCRIPTION OF THE SPIKE PROTEIN CAME IN 1972, IN A paper describing OC43, that now-harmless human pathogen that likely caused the Russian flu over a century ago. It wasn't much of a portrait, though. The authors of the paper, despite using electron microscopes to reveal the spikes, which appeared like bright, stubby tentacles reaching out for purchase amid a void, couldn't explain anything about them. "Repeated attempts to determine the composition of the projection antigen," they wrote, "were of limited value." The spikes were there, no doubt about it, looking like blurry white pinpricks of light barely touching the virus's protein shell, but what they were made of and what they did was beyond the reaches of the scientific tools of the day. A decade later, scientists had begun to decipher a few more tentative clues about the spike: it was made up of two separate regions, dubbed S1 and S2, which worked together in different ways to allow coronaviruses to latch on to and invade cells.

S1, the tip of the spike protein, contains the receptor-binding domain, so called because coronaviruses use it to bind to the cell walls of their hosts. Once fixed in place, the spike's stalk (the S2 region, which contains the fusion membrane) then unfurls itself and goes to work dismantling the cell's defenses. The spike works as a one-two punch, with S1 latching the virus on to the surface of a cell like a grappling hook, and S2 then fusing with the cell walls to create an opening through which the virus can enter, take control, and copy itself.

While the broad functions of the spike protein were known in the 1980s, nobody could explain how they allowed coronaviruses to infect multiple hosts so effectively. Baric, in a major breakthrough, discovered that the S1 region, the viral hook, was elegantly adapted to latch on to the cells of a vast number of species. And this is where the virus's molecular makeup hints at a more complex story. S1 is perfectly designed to fit specific cellular receptors known as orthologs, which are genes expressed identically across species that share common ancestors. Over 90 percent of mice genes, for example, have human orthologs, humbling evidence of our shared ancestry. For viruses inventive enough to attack them, orthologs can act as a shortcut to attack all manner of creatures.

This streamlined attack on orthologs, though, raised a very disturbing possibility about the nature of our relationship with coronaviruses. Baric realized that it was almost a certainty that coronaviruses evolved to infect an ancestor common to humans and other animals susceptible to their attack. And this meant that these pathogens have been evolving for eons—by some estimates, for as long as three hundred million years.

Picture, then, a coronavirus hunting not only early humans but also the primeval creatures that predated us and from which both we and our rodent kin descended. Though they revealed themselves as a pathogenic threat only in the twenty-first century, coronaviruses are no new enemy. They are, instead, a family of pathogens that have been honing their deadliness since before humanity was born and that long ago equipped themselves with a skeleton key that could unlock our defenses.

MUTATION HAPPENS IN TWO WAYS. THE FIRST IS FOR DIFFERENT NUcleotides to be mistakenly inserted when viruses replicate; these small changes, which can be as minor as a single nucleotide among thirty thousand, are called antigenic *drift* because the variant virus's genome has drifted ever so slightly away from its progenitor. The second way mutation happens is for whole sections of genetic code (blocks containing thousands of nucleotides) to be swapped out with sections of a different variant or virus that also happens to be replicating in the same cell. This is called antigenic *shift*, given how alarmingly different the progeny virus can be from the virion that spawned it.

Antigenic drift happens often, and it's unlikely to dramatically change the basic approach a virus uses to infect a host. Antigenic shift, however, is orders of magnitude rarer, requiring as it does that multiple viruses replicate in the same cell at exactly the same time. In small and relatively stable viral families, this almost never happens, making antigenic shift sporadic at best. But coronaviruses, which Baric found were always "drifting" just under the threshold of error catastrophe, produced plentiful variants and new strains all the time, many of which could also move freely across multiple species because of the architecture of their spike proteins. This greatly upped the chances that different coronaviruses would find themselves infecting the same cell. And this close contact increased the speed at which abrupt and profound changes (antigenic shifts, also known as genetic reassortment) would take place.

The spike protein unlocked entry into the cells of a wide range of species. The reason, Baric found, was because antigenic shifts (the wholesale swapping of genome sections) occurred frequently in the S1 region of the spike protein, the part that looks like a budding flower and that the virus uses to hook itself on to cell walls. When Baric tracked them, he found that antigenic shifts happened three times more frequently in the S1 region than in any of the other components that made up coronaviruses. The S1 region wasn't a skeleton key, per se: the *Coronaviridae* were liberally trading keys among themselves until they landed on the one that fit the cell they were trying to pry open. It was yet another elegant strategy the family had unleashed to propagate itself across time, space, and species. And it had made coronaviruses a link, at last count, to creatures as varied as bats, mice, minks, orangutans, camels, pangolins, civet cats, cows, ferrets, pheasants, beluga whales, cheetahs, and humans, the list of species in the chain growing ever longer.

2003, Chapel Hill, North Carolina

When the SARS epidemic hit, Baric was more shocked than anyone, but not because he hadn't seen it coming. His experiments with MHV had revealed that coronaviruses were masters of interspecies transmission, and by 2003 it had been almost a decade since he first sounded the alarm that these strange, too-large viruses should be considered emerging threats. Like everyone else, though, as the outbreak of deadly pneumonia first spread across Southern China and Hong Kong, Baric had no clue what pathogen was causing it. Nevertheless, deep in the back of his mind, he allowed himself the thought that a coronavirus might just be the culprit.

Then, in the early months of 2003, he got a phone call from a contact at the U.S. Centers for Disease Control and Prevention, who showed him images of the virus first isolated by Malik Peiris in Hong Kong. Baric, rapt, stared at the images and traced the familiar pinpricks of light surrounding the ponderous circular mass. There was only one possible explanation.

"It was consistent with what we were seeing in animal populations," he said of the SARS virus. "It was consistent with what could happen rapidly in cell culture." Still, it was chilling. And as Baric learned more about the new pathogen, he became more disturbed. In many ways, SARS had many of the hallmarks of the laboratory coronaviruses, like MHV, he had been studying for two decades. But in other ways, it was entirely, terrifyingly different. "This virus . . ." he said. "I mean, there was no coronavirus like it." SARS wasn't some predictable variant of a well-known suborder of coronaviruses. Its genetic sequence placed it at the furthest known edge of the family, in a theoretical subunit called Group 2 that should, in principle, have been the missing link between the other three groups of coronaviruses. Until SARS emerged, Baric explained, "nobody had any evidence that Group Two viruses even existed, to tell the truth. So, that was a shock."

It didn't end there. It was one thing to have theorized about an emerging viral threat, but another thing altogether to see your theory come to life in the real world, where it would infect over eight thousand people and end up killing almost 10 percent of them. "Working on emerging viruses is a world of mixed emotions," Baric said. "Nobody wants to see a new virus emerge and cause morbidity and mortality and death and human suffering, regardless of where it is in the world. But at the same time, it's a new member of a virus family that you've worked on for decades, and you have opportunities to contribute to the human response." It was, he said, "an exhilarating kind of feeling, with a sickness in the pit of your stomach."

There is a sanctity in lab work that derives from its solitude and uninterrupted rigor; in this and other ways, it is not unlike monastic life. A band of adherents is brought together by a deep fervor for mystical truths visible only to a select few. And if the first Christian monks were brought together as a *koinonia* (communion with others seeking universal truths), then modern-day laboratories, where fundamental laws are revealed and alien worlds are probed, are where the passionate go for revelation mediated through the advanced tools of discovery: the electron microscope, the western blot, the X-ray diffractometer. Unlike monks, who submit themselves completely to God's will, scientists reveal truths by exerting absolute control. Without that control, the edifice falls to pieces. When Baric, who had always sought out places of isolation, immersion, and control, was confronted with the news that SARS had entered human populations, the barrier between the holy sanctity of his lab and the world that surrounded it crumbled.

SARS brought Baric from the outer fringes of abstract virology and squarely into the center of epidemic science. Until the first case of atypical pneumonia was reported in China's Guangdong Province in November 2002, the coronaviruses that Baric manipulated so freely in his experiments were nothing more than his subjects and tools. Regardless of the bounds beyond which he pushed the virus each day, when the fluorescent lights snapped off and the infected tissue was placed back in storage within industrial coolers, it was his property. SARS inverted the dynamic. With the virus loose in the world, he could no longer claim to be above the fray. He became, instead, like the rest of us: a susceptible host. It was a jarring shift, but this descent from abstraction would also make him one of the most consequential scientists alive.

Chapter 4

IN THAT CASE, YOU CAN GO TOMORROW

Monday, May 5, 2003, New York City

A T THE HEIGHT OF THE SARS EPIDEMIC, DR. W. IAN LIPKIN returned to his office at Columbia University to find a Chinese consular official patiently waiting for him. Lipkin had just finished giving a lecture to his Ph.D. students about a new technique he had devised to accurately identify the deadly new coronavirus, and he felt confident—it was a familiar feeling—that it might just be the key to ending the epidemic. He just didn't know that word would travel so fast.

Lipkin was the director of Columbia University's Infectious Disease Laboratory, and a scientist with a reputation for relentlessly attacking big scientific problems no matter how colossally difficult. At the time, the SARS epidemic was, though slow-moving, proving unstoppable for a simple reason: the first-generation methods that had been developed to test for it were error-prone, making it hard to know who really had the virus and who didn't. A reliable diagnostic tool, Lipkin knew, could change all that. If public health officials could be sure of who was infected, they could rapidly extinguish outbreaks when they flared by immediately imposing quarantines, lockdowns, and the other necessary steps required to control the spread of infectious disease.

But any old test wasn't enough. A test that was too broad would pick up many different viruses, including the other endemic human coronaviruses; while one that was too specific might be too finicky to pick up the virus at all. For a test to be useful, it had to be reliable and sensitive enough to accurately detect whether the SARS virus was present even in someone with an extremely low viral load. This was, using the technology of the time, no small feat.

Before SARS came around, Lipkin had made his name in scientific circles by pioneering a radical new approach to mapping viruses. Eschewing the microscopic universe of whole viruses, he instead homed in on the constellation of molecules that made up the building blocks of all life. In the early 1990s, he was the first to reverse-engineer the entire genome of a virus by painstakingly collecting infinitesimally small molecular fragments culled from animal tissue and then stitching them together based on his prior knowledge of which sections went where to create a coherent viral whole. The task was as delicate and time-consuming as it sounds, and it demanded a rare kind of patience. It took Lipkin over two years to clone the three-kilobase-long genome of the bornavirus, which was largely ignored because it didn't pose a threat to humans, instead attacking the central nervous system of animals. (A little over a decade later, Marra and his team were able to sequence the entire thirty kilobases, thirty thousand nucleotides, of the SARS genome in a matter of days because computing power had advanced sufficiently to automate the process.) Lipkin was the first to use purely molecular methods to identify pathogens, an approach we now take for granted. It also established him as one of the world's best epidemiologic toolmakers: a scientist with a seemingly unending capacity to develop techniques to meet the crises of the moment. Lipkin evidently felt that the contribution hadn't received its due. He had, after all, created an entirely new field of molecular virology, without which we would have been largely unable to rapidly identify coronaviruses and other viral threats. "But they say," he stated wryly, "who gives a shit."

With his hawklike face, flop of brown hair, and piercing eyes,

Lipkin exuded a kind of spatial intelligence, as if he were reading the room and everyone in it for signs of weakness and strength. He was a small man and possessed a natural swagger, moving animatedly and with a kind of studied staging, his public gestures and comportment seemingly designed to invoke confidence and trust. In private, he would uncoil, forcing his interlocutors to keep up with him as his thoughts tumbled quickly into the deeply technical, a place where he appeared to feel most at ease and also most alone.

Lipkin had watched from afar as the novel coronavirus spread, slowly but unrelentingly, across Southern China, Hong Kong, and then the world. Though it became clear that SARS was largely transmitted within hospitals (a phenomenon called nosocomial transmission) rather than in the community, he also knew that its spread in public spaces wasn't entirely impossible. One of the inflection points early in the epidemic was the decision by a single physician, Dr. Liu Jianlun, in February 2003, to stay on the ninth floor of Hong Kong's Metropole Hotel after having unknowingly acquired SARS from a patient elsewhere in Southern China. Liu's one-night stay at the hotel ended up infecting seven other people on his floor; one of these was the Canadian tourist who then seeded the Toronto epidemic, which ultimately killed forty-four people, including her and her son. A Chinese American businessman staying on the same floor that night perished two weeks later in a Hanoi hospital after launching an outbreak in Vietnam. Then, in April 2003, a month prior to the Chinese consular official's visit to Lipkin's office, Hong Kong experienced its largest community outbreak, at the Amoy Gardens complex, a vast 1920s-era industrial space that had been converted into a residential facility housing more than ten thousand people. The building's poorly designed drainage system had allowed fecal matter containing SARS to become aerosolized and drawn by ventilator fans into tenants' apartments. More than three hundred people subsequently became infected.

With existing testing regimens and public health measures failing to slow the virus, Lipkin immediately saw a need, and an opening, to adapt his molecular toolbox to the detection of the new pathogen. Starting with the genome that Marra and Brunham had gifted to the world, Lipkin set out to develop a reliable diagnostic test. At the time, the first assays to detect SARS had been rapidly designed by Malik Peiris, and while they were an effective frontline defense, Peiris had been forced to rush them into the field without fully validating them, and they didn't always pick up the virus in infected people. They also only worked days after someone's initial infection, so by the time an infected person was positively ID'd, they could already have spent days spreading SARS far and wide. Lipkin figured he could do better.

The first step was to use PCR (polymerase chain reaction) to detect whether the virus was present in the blood, even at very low concentrations. (This was the same technique Marra used to amplify the minuscule amounts of SARS virus in the TOR2 sample.) A PCR test uses "primers," short strands of cloned DNA that have to be specifically designed to match and then attach to sections of a viral genome. Luckily, by May 2003, Marra and Brunham's SARS genome sequence had been widely shared, so Lipkin could search it like a map for the regions unique to SARS, meaning his test wouldn't mistake it for benign human coronaviruses like OC43.

Lipkin was well aware of just how valuable his test could be in turning the tide of the nascent epidemic. That's why, when he presented it to a group of Ph.D. students, he was rankled that some of them didn't seem all that interested. Two students in particular (both visiting from China) kept interrupting his presentation. After running through the sensitivity of his diagnostic test, first one and then the other got up and left the room. They soon came back and then, just as soon, left again. With only eight or ten people in the room, the constant back-and-forth was impossible for Lipkin to ignore.

After the lecture, he strolled back to his office, where he spied a man waiting outside the door. As Lipkin approached, the man got up and introduced himself: he was from the Chinese consulate and had just heard about Lipkin's work. The Chinese students, it appeared, had left the room during his lecture to alert their contacts about a potential new weapon in the fight against SARS. After some brief pleasantries, the two exchanged contact information, and the Chinese official bowed and left. The following day, Lipkin received an invitation to tea at the consulate, which he dutifully accepted. In a meeting room inside an imposing concrete-and-glass structure overlooking the Hudson River, four Chinese government officials quizzed him gently about the SARS test he had been developing. Lipkin understood the stakes: China was fighting the epidemic blind, with no clear notion of how, or how far, the virus had spread. The consular officials, seemingly satisfied with his answers, thanked him for his time. Assuming that this was the end of it, Lipkin returned to his office.

But the next day, an invitation arrived: Would Lipkin attend dinner that night? Evidently there was more to discuss. That evening, he arrived at the appointed spot, a nondescript Chinese restaurant that had been hastily divided by screens into separate rooms. Quickly ushered to the back, he was brought to a table teeming with plates of exotic food—abalone, shark fin soup, "and things that I really shouldn't be eating, because they're all endangered," he recalled. Seated at the table were the consular officials who had been his companions over the last two days, and who were obviously impatient to get down to business. After the requisite formalities had been dispensed with, they came to the point:

"We need you to go to Beijing tonight."

"I can't go to Beijing tonight," Lipkin replied flatly.

The officials were unsatisfied; how could it be that one of the world's preeminent epidemiologists had better things to do than help end a deadly pathogenic threat?

"First off," Lipkin replied, "this is the first I'm hearing of it. Second, I don't have a visa. And I also can't get permission to go because there's a state advisory against travel to China." The officials considered his reasons for a moment and nodded thoughtfully. "In that case," they told him, "you can go tomorrow."

For all the absurdity and gravitas of the situation, Lipkin was unfazed. "I was not surprised," he said. "When you have a tool kit, people come to you for help." And as far as he could tell, China needed him more than he needed it. The country had both a scientific problem and a political one. The scientific problem was the SARS virus, which would continue to spread until it could be accurately detected. The political problem was just as confounding: The Chinese government was seeking to reintegrate the country into the global economy right at the moment that a coronavirus had severed it from the rest of the world. The epidemic only added to the deep internal fissures that the planned economic integration had engendered.

"This was a period when there were Maoists in the street," Lipkin said, referring to the political faction protesting the government's efforts to Westernize parts of China's economy. And despite the fact that China was ruled by a dictatorship, the reach of the central government was still relatively weak. The Chinese president, Jiang Zemin, had a time-limited term and was forced to manage the many different factions of the country's Communist Party apparatus, including its ardent Maoists, in order to remain in power.

Lipkin figured he could solve both their problems. By arming the Chinese health system with reliable SARS tests, he could help the government finally get ahead of the epidemic and stabilize the country's economy. Bringing in a Westerner would help smooth relations with outside agencies (like the World Health Organization) that were proving vexatious to the Chinese state. Still, Lipkin wasn't keen on going. Travel to China had practically ceased during the epidemic, and governments were urging their citizens to stay away. But after failing to find colleagues at other institutions willing to make the journey—"They were all told that if they go, they'd lose their tenure; nobody wants anybody coming back with SARS"—he relented. After all, somebody had to play the hero. So, Lipkin flew to China. On the last leg, from Japan to Beijing, the only passengers were him, his colleague Thomas Briese, and Elisabeth Rosenthal, China bureau chief for *The New York Times*, who was heading into the heart of the epidemic to retrieve her children and bring them back to the United States.

Once Lipkin arrived, the first order of business was brokering a peace between the World Health Organization and the Chinese government. After the SARS outbreak, the WHO's China country lead (the same man who had phoned Malik Peiris on Valentine's Day) had become incensed by the lack of information he was provided about the severity of the novel epidemic. In a fit of rage, he had managed to alienate the national government to the point that the lines of communication between China and the WHO had completely broken down. "He was just screaming at them all the time about how they were hiding information," said Lipkin, dumbfounded by the man's shortsightedness. Screaming wasn't going to bring the Chinese government back to the table. Lipkin, the son of two psychotherapists, preferred a more conciliatory approach, one that kept conflict to a minimum and, in the style of his hosts, put a premium on backroom negotiations rather than public condemnation. It worked; Lipkin was able to mollify China's public health leaders and renew their commitment to coordinating their anti-SARS efforts with those of the WHO.

While his diplomatic machinations were useful in keeping a lifeline between the Chinese government and the international community, it was Lipkin's PCR test that changed the game. He had hand-carried ten thousand tests from New York and had been instructed to train Chinese public health workers on how to use them. Part of that training, he was told when he arrived, would involve a live appearance on Chinese television to demonstrate how the test worked. When he got to the TV studio, Lipkin saw that the thermal cycler for the PCR test (the machine that analyzed and reported the results) had been hooked up to a projector, which was broadcasting across four live television feeds. The entire country was going to bear witness to whether this new SARS test worked, in excruciating detail. Lipkin was told to stand at a table across from four military generals wearing uniforms covered in voluminous ribbons and shiny medals he couldn't stop staring at. If Lipkin was worried about the reliability of his test, this moment—when it was being watched by China's military brass and beamed onto television screens across the nation—was as good as any to prove it worked.

Lipkin introduced a sample of virus onto his bespoke assay, and the thermal cycler hummed to life. Then he waited as the instrument initiated the chain reaction that would, in principle, lock on to RNA fragments from the novel coronavirus and force them to do what they did best: replicate over and over, to the tune of millions upon millions of fragments, until the mass of genomic material grew so large that it could be detected by the test. Until the signal flared, Lipkin could only wait and hope, acutely aware of the gaze of the camera lenses and the eyes of the generals trained on him. After a few excruciating minutes of nothing, the screen finally lit up: on the sixteenth cycle of the chain reaction, the test finally detected SARS.

After the successful demo, Lipkin's tests were distributed across China's epidemic front lines, which were located, with rare exception, inside hospital wards, where SARS patients were unknowingly infecting their caregivers. With the PCR test in hand and improved public health measures in play, the number of healthcare workers getting infected dropped precipitously: the movement of the virus from person to person could finally be accurately tracked, and those infected could safely be isolated, leaving the virus with dwindling options for transmission.

The success of Lipkin's test was a bellwether marking the moment when the epidemic began its retreat. But it was also something more: it was evidence that, for all the fear and consternation that had arisen with SARS, humanity was equal to the challenge. A deadly coronavirus had surprised everyone (with, perhaps, the exception of Ralph Baric), and yet, for all its guile, the family's pathogenic offspring could be chased, hounded, and forced back into the wild. The science of the moment bested the coronavirus that had emerged to test us. It was, in retrospect, a false victory.

Beyond its scientific applications, Lipkin's test had other, less immediate consequences. His trip to China in 2003 would mark the start of a relationship spanning decades. "I wound up getting a medal," Lipkin said with mordant bemusement.

Lipkin became a scientific emissary of sorts between the Western world and China. He helped establish a research group in

Guangzhou, in the cradle of the SARS epidemic, which was focused on emerging pathogens, and he spent months and then years traveling back and forth between New York City and Beijing to meet with dignitaries, public health officials, and senior members of the Chinese government. It was a model of cooperation largely removed from the political machinations that normally dogged relations between East and West, and it pointed the way forward for science to be the tip of the spear in the international community's response to epidemics, wherever they might arise.

But as he reflected on the success of the fight against SARS, Lipkin was more preoccupied with the scientific strategies that had been forgotten than with celebrating what had been accomplished. Counting Malik Peiris as a friend, Lipkin was one of the few who truly understood the tenacity and cleverness that had been required of Peiris to first isolate SARS amid all the panic and noise that pointed to a new form of avian flu. And while the world at large thought of genomics as a moneymaking experiment to map the entire human genome, Lipkin knew just how stacked the odds were against Brunham and Marra's sequencing the novel coronavirus; in his words, what they had accomplished was a "heroic effort." And yet, all that mattered to most people was that the strange new virus was far away and moving so slowly that it would probably just burn itself out. "Only a few people remember how difficult it was to do this work," Lipkin said caustically. "Only a few people really appreciate what was required."

Lipkin became dismayed that the world had learned exactly the wrong lessons from SARS. The ostensible ease with which the novel coronavirus was becoming contained through testing, quarantine, and public health measures left most observers to draw mistaken conclusions. The first was that rapidly containing epidemics was the norm. From the outside, Brunham and Marra's herculean efforts to map the SARS genome and Lipkin's rapid development of a PCR test exquisitely matched to the molecular structure of the virus appeared to be the kind of science that just happened and that would save the day every time a new pathogenic threat emerged. The second mistaken conclusion was that coronaviruses were, by and large, slow-moving pathogens and, SARS notwithstanding, largely innocuous. Lipkin knew better.

In the mid-1980s, two decades before his career would intersect with SARS, Lipkin was working at the Scripps Research Institute in San Diego. At the time, some of his colleagues had been running tests on MHV, the mouse coronavirus that had been a stalwart of laboratory research since the late 1970s. "One of the things that I remember from that time," Lipkin said, "was that when you put MHV into a cell colony, it went through it like wildfire. There was always an enormous problem with contamination." The problem was that MHV was so adaptable that, if given the chance, it was liable to infect different cell lines across entire laboratories, regardless of what kinds of animal tissue were present; it was Baric's experiment with hamster cells spun out of control. "We had the same problem when I moved to Columbia University," Lipkin added. "A whole animal facility was shut down because MHV was so contagious." It was a portent of what was to come.

Chapter 5

WE NEEDED ANOTHER
$108 MILLION

Throughout the entirety of the SARS epidemic, the United States recorded a total of twenty-seven cases and zero deaths. SARS was, at most, one of those scary news stories about near-impossible events that happen to other people in other places that look nothing like our own. It was easy to assume that an American SARS epidemic didn't happen because the country was somehow better prepared to avoid it than, say, China or Vietnam. This narrative fit neatly with the way most Americans perceived their healthcare system: The United States was the most scientifically innovative country on earth, with an unmatched capacity to find medical solutions to meet the moment. It was the home of HIV triple therapy; of futuristic cancer treatments; of vaccines for polio, measles, and the flu. If a problem emerged, the American scientific establishment would solve it. And if other countries experienced outbreaks, it was because they didn't have the scientific prowess required to beat back these aberrations in nature.

But it was the failure of SARS to gain a foothold in the United States, and the assumption that American science would always find an answer, that set the conditions for a pandemic to take hold. The original trauma of SARS was never visited upon the United States; the scar tissue never formed. The country, like many others, remained as vulnerable as ever while believing itself ready for whatever monstrosity nature next produced. It was hubris of the highest order.

June 2003, Hong Kong, People's Republic of China

By the summer of 2003, the genetic code that Brunham and Marra mapped became an invitation to power and influence. Though Lipkin's PCR test had been brought out into the field and had helped prevent the epidemic from spreading farther, SARS tenaciously resisted eradication in Hong Kong, where the case fatality rate had, shockingly, reached 15 percent. The government of Hong Kong needed a lifeline to chart a path out of the epidemic. In Brunham's commitment to genomics in service of public health, they believed they might have found one.

One month after Lipkin's trip to Beijing, Brunham was flown first class on a Cathay Pacific flight to Hong Kong, where he arrived well rested and amply fed after the mostly empty flight. Descending a mobile staircase onto the tarmac in June 2003, he was greeted by a coterie of Hong Kong government officials who, after warmly welcoming him to the territory, escorted him to a waiting limousine. From there, he was whisked to the Grand Hyatt Hong Kong, where, though it was past midnight and the lobby was largely empty, he found himself in good company. At that late hour, the only other guests checking in were the Rolling Stones, who had come to Hong Kong for a series of planned shows to highlight the city's recovery from the epidemic. After a long career spent in scientific obscurity, Brunham was finally center stage. He had become a rock star in his own right, practically overnight.

To hear him tell it, his ten-day trip to Hong Kong was partly ritualized thanks—as if he were a knight-errant receiving the keys to a city he had helped save—and partly a beleaguered SOS. Between helicopter rides, anodyne meetings with government officials, and his time ensconced in the opulent Grand Hyatt, Brunham was also there to share all he knew about the power of genomics. One of the scientists intrigued by his work was Malik Peiris, who invited Brunham to visit his lab at the University of Hong Kong. There, Brunham was shown vast enclosures of ducks, rodents, and other wild animals, which Peiris had used to first isolate SARS and was now using to track the evolving epidemic.

Brunham came armed with his own agenda. Ever since he and Marra mapped SARS, he had been clinging to one burning desire: he wanted to create a vaccine so that healthcare workers and their patients infected with the novel coronavirus would stop dying. It wasn't just professional; some of Brunham's close colleagues in Toronto had been infected with the coronavirus, and one had died. But he also wasn't looking for personal gain. Despite his veneration of J. Craig Venter's accomplishments in genomics during the time they collaborated, Venter, Brunham said, "wasn't a role model for my view of how science fits into the larger human experience." Brunham wasn't just uninterested in science for profit; he was an ardent supporter of open science, a model in which all scientific knowledge was freely shared and in which no discoveries were patented. The single-minded way that the Hong Kong and Chinese governments were working to eradicate SARS without an apparent interest in monetizing the science deeply intrigued him. Here were highly motivated partners, he figured, who understood the need for a vaccine, who were likely receptive to an open-science framework for developing one, and who could play a key role in an international vaccine development consortium.

After being wined and dined by China's elite in Hong Kong and Guangzhou in the aftermath of his team's sequencing of the SARS genome, Bob Brunham returned to Vancouver feeling empowered. He was at the apex of his scientific influence, and he knew it: it wasn't every day that you beat the U.S. Centers for Disease Control and Prevention, the world leader in infectious disease prevention, at its own game. He and Marra had taken five nanograms of virus and transformed them into an unprecedented scientific achievement:

the mapping of a new and deadly coronavirus. He now had the reputation, the clout, and the connections to make good on the promise of his entire career in vaccinology.

A vaccine to end SARS would have been enough. But on his dreamier days, Brunham allowed himself to imagine where that single successful vaccine could take humanity. If he could develop a vaccine using a nonprofit, collaborative model, then he wouldn't have just bested the novel coronavirus. He would have shown the world a new way of confronting its enemies, one unfettered by the backward-facing logic of the free market, by looking to the future. What's more, if he were able to make one vaccine against a coronavirus, that would turn the development of vaccines for any problematic future *Coronaviridae* family members into what he described as a "plug-and-play-type process"—the infrastructure would be in place, and the vaccine platform's safety record would have been proven, so it could be as simple as swapping out one virus for another.

It was an audacious play, but one that was sorely needed to reverse the descent of the field. In the past fifty years, only one successful vaccine had been brought to market. Like practically all other areas of scientific innovation, vaccine development had taken a nosedive in the second half of the twentieth century as the costs of innovating climbed higher and higher. It wasn't that the scientific work wasn't continuing—Brunham himself published more than seventy papers outlining vaccine mechanisms and candidates between 1983 and 2003—it was that there was nowhere for the discoveries to go. Pharmaceutical investment in novel vaccine development had come to a standstill because the risks were just too damn high. Why throw tens of millions of dollars at a clinical trial of an unproven new vaccine when an incrementally better chemotherapy drug was a surefire way to make money? While the biggest pharma companies made astronomical profits, they also, like many massive companies, continually owed debts to their lenders. The pharma giant Pfizer generally owed almost twice its annual revenue, due over ensuing decades, which effectively locked it into the same way of doing business it had always employed. This model, built on humanity's predictable need for the same old cures—from pain, from cancer, from strokes, from creeping decrepitude—had been refined over a century. The market was not designed to take chances on vaccines for novel viruses.

Brunham figured that SARS could disrupt this paradigm. In the early months of 2003, when the virus was harvesting death across the globe, it seemed more likely than not that a vaccine would be the only way to stop it. But because he had spent his career moving among varying epidemic threats (the killer fungus *C. gattii*, chlamydia, HIV, syphilis), he also understood that what humanity needed wasn't just a SARS vaccine but a commitment to vaccine development writ large. There were unchecked diseases killing humans in the hundreds of thousands each year, all of which could be eradicated with vaccines. SARS was just the foot in the door. Armed with scientific credibility, gravitas, and vision, Brunham made his play.

First, he launched the SARS Accelerated Vaccine Initiative, or SAVI (pronounced like *savvy*), with great fanfare in April 2003, within weeks of the sequencing of SARS. SAVI's goal was to distill the making of vaccines—which generally took at least a decade from animal models to commercial product—into a process that, from start to finish, would take less than two years. It was risky, audacious, and near impossible, but neither Brunham nor his collaborators cared about the odds. They just wanted to make a SARS vaccine, and fast.

SAVI was a network of labs that all agreed to test vaccine candidates but to share in the prize equally, no matter which specific lab actually hit on a vaccine that worked. Once a viable cure was discovered, a group of pharmaceutical partners would be standing at the ready to take SAVI's laboratory research and commercialize it. The idea was appealingly simple and motivated by an antiquated ethic that held that the only way to stop a pathogen from killing people was to work together to make a vaccine available to anyone who needed it as cheaply as humanly possible. It was, Brunham said, "a road map to making a vaccine for the public good," a concept that had fallen by the wayside as the twentieth century, with the global dominance of increasingly unshackled capitalism, came to a close. SAVI, Brunham thought, could change all that.

Evidently, others thought so, too. Vancouver was experiencing its own SARS outbreak (seeded by a patient who had stayed at the Amoy Gardens complex in Hong Kong), and local politicians had turned to scientific leaders like Brunham for guidance. The answer Brunham gave them was that the best way to keep people safe was a vaccine and that the best way to make one was, of course, SAVI. It was hard to quibble with the logic. By the end of April, Brunham had received a two-million-dollar grant from the provincial government to kick-start the project. The next step was to sell SAVI internationally.

The first hiccups came on that fateful trip in June 2003, when Brunham visited Hong Kong and Guangzhou at the behest of the Chinese government. He was treated like scientific royalty, and when he unveiled his pitch for SAVI, the assembled government officials and scientists appeared keen. In a major win, he was even able to get the Guangdong Provincial Center for Disease Control and Prevention to agree to run human clinical trials as soon as SAVI landed on a viable vaccine candidate. There were also plans for a joint Chinese-Canadian SARS conference that would be held in Guangdong Province in November, to make sure that neither side lost momentum.

But when it came to funding, Brunham found no takers. He was disappointed, but he understood the rationale. By June 2003, traditional public health measures, including rapid testing, quarantining, and restrictions on movement, were working to put the SARS genie back in the bottle. So, why would the Chinese government spend millions funding a program that it almost certainly wouldn't need? Brunham came home empty-handed.

Back in North America, he faced even stiffer headwinds. At least the Chinese understood the full gravity of SARS. The hundreds dead and thousands infected had etched the virus deeply into the country's national psyche. In the United States, however, where no

one had died and where only a couple dozen people had been infected, a SARS vaccine was a laughable investment. As Brunham
well knew, his two-million-dollar grant was a drop in the bucket.
Running preclinical animal models of vaccine candidates alone generally cost upward of ten million. To then get a viable vaccine candidate biomanufactured and tested in Phase 1 (safety) and Phase 2
(efficacy) human trials would cost at least another one hundred million. Doing the math, Brunham couldn't help but laugh. "We had
two million, and we needed another one hundred and eight million."
In the United States, zero deaths meant that the added value of a
SARS vaccine was also zero. Brunham and Marra were recognized
by their scientific peers the world over for their mapping of the
genome—Lipkin praised that effort effusively for its heroism, and
Baric offered them praise in the most laudatory form he knew: by
repeatedly citing their work and using the SARS sequence they had
mapped to power experiments that stretched our knowledge of what
mutations made coronaviruses deadly—but beyond the hardcore
coronavirologists, there was a noticeable lack of interest among
American research institutions in Brunham's pitch.

Despite being entirely shut out of major funding sources, and
with the SARS epidemic waning across the world, Brunham still
had one last trick up his sleeve. By the summer of 2003, within a
few months of its launch, SAVI had started animal tests on three
vaccine candidates—a killed vaccine made of dead SARS virus, an
adenovirus-delivered SARS vaccine, and a vaccine containing the
SARS spike protein—that showed real promise. By triaging the
work to various labs across the nascent SAVI network, Brunham
and his collaborator Brett Finlay had in just six months generated
solid evidence for the efficacy of their vaccines, which marked a
major milestone before testing commenced in humans. In 2003, this
kind of speed was unheard of: it usually took close to five years to
identify the appropriate antigen (the part of a virus the body's immune system targets) and to identify specific antibodies in animals.
Against all odds, and even without major backers, Brunham's vision of an open-science approach to vaccine making was coming to

pass. Others were evidently watching: Ralph Baric had been steadily keeping tabs on SARS vaccine candidates and had been increasingly citing SAVI experiments in his papers; he had even used Brunham and Finlay's work as a jumping-off point for his own increasing interest in vaccine making.

The next step was human trials. And for these, Brunham and Finlay needed one thing: SARS-infected people on whom to test their vaccines. While SAVI had been launched just as the first wave of SARS was starting to slow, both Brunham and Finlay knew that respiratory disease epidemics almost always had a second wave, often arriving in the fall, that was generally far worse than the first. When cases started rising again, Brunham was sure that funders (pharmaceutical companies and government) would finally see the light.

As 2003 came to a close, Brunham waited for the inevitable onslaught of SARS infections. SAVI's agreement with the Guangdong Provincial Center for Disease Control and Prevention provided them with the ideal site to conduct human trials: the place where SARS first rose out of the abyss. It was just a matter of waiting for the virus to reemerge.

But the second wave never arrived.

As the SARS outbreak expanded, stabilized, and spread across the globe, Ralph Baric tracked it as it went, collecting samples from those creatures, both human and animal, that had become infected along the way. Where the world at large experienced the epidemic as a blur, Baric saw, with the samples as his guide, three distinct phases, each driven by key mutations that made the virus better adapted to take advantage of its circumstances. During the first wave, the proto-SARS virus was only one of many coronavirus strains that had found their way into humans and were being challenged to replicate successfully in their new hosts. While most strains could not survive the leap, the proto-SARS did, replicating efficiently enough in the respiratory tract of a single person to lay the groundwork for an epidemic.

The second phase, according to Baric, started in a hospital. That was where, through antigenic shift, the proto-SARS recombined with other coronavirus strains that happened to be circulating in an individual and, in the process, picked up key mutations that let it not only replicate efficiently inside one human being, but transmit itself to others as well. This was the moment Baric zeroed in on as the true start of the epidemic.

By analyzing samples from SARS-infected patients later in the epidemic, he pinpointed the beginning of the third phase as the moment when SARS went global. That was, he discovered, the result of the virus undergoing further chance mutations—antigenic drift—that transformed it from a slow-moving pathogen into a virus efficient enough to spread rapidly across human populations. This heightened efficiency would have come, Baric knew, from chance alterations in the virus's spike protein that would have allowed it to tighten its grip and more easily enter cells, or better conceal itself from the human immune system's roving sentries. It was this last SARS variant that allowed the virus to fly across the world, spreading from places like the Metropole Hotel to the faraway metropolises of Taipei, Hanoi, Bangkok, Manila, Kuala Lumpur, Sydney, Toronto, and Vancouver.

Through experimentation in his lab, Baric had discovered that the rate at which coronaviruses mutate speeds up when they're forced to switch between host species. This was exactly the situation that would have occurred in the first phase of SARS, when the proto-SARS virus jumped from an animal to a human host. Once this initial adaptation occurred, the rate of mutation slowed down—but it never stopped completely.

Baric recognized that environmental circumstances (hospital staff without personal protective equipment; international air travel connecting distant locales) were critically important in spreading SARS. But he was able to map how each environment the virus found itself in pressured the virus to mutate into a more refined form. As he tracked this trajectory, he believed that it would have been just a matter of time before SARS evolved even further to initiate a fourth phase: a full-blown pandemic.

This didn't happen, of course. And the reason, Baric believed, wasn't because of human ingenuity. It was, he said, "because we got lucky." SARS didn't become a pandemic threat because of a key flaw in its design: people were most infectious—that is, they had the highest amount of viral copies in their system and were shedding them liberally—when their symptoms were the most severe. This alignment between the peak of a host's symptomatic period and their infectious period was humanity's saving grace, because people were generally isolated in hospital rooms when they were most at risk of spreading the virus to others. Even better, that peak period lasted for only about thirty-six hours. "If the virus had transmitted more efficiently," Baric explained, "even thirty-six hours before the onset of symptoms, we never would have stopped it. Because there would have been community spread, asymptomatic spread, and it would have been SARS-2." Humanity, Baric believed, had been saved not by a flawless public health response but by a flawed pathogen.

While others celebrated having reined in the epidemic, Baric couldn't help shake the feeling that it was a hollow victory. For all the damage it caused, SARS was poorly adapted to spreading among humans; it wasn't even among the worst-case scenarios that Baric had experimentally evolved in his lab. In fact, for all the trauma SARS caused across the globe, the virus's short tenure among our species made it by far the worst-adapted of any of the known human coronaviruses.

The end of SARS brought a close to a terrifying epidemic, but it also marked the beginning of an intense new phase in Baric's career. His was one of a handful of laboratories around the world that kept SARS samples after the epidemic was over. It wasn't because he saw the virus as a relic of a defeated epidemic. With the shock of the SARS epidemic fresh in his mind, and unable to set aside what he knew about the tenacious adaptiveness of the *Coronaviridae,* Baric could not simply move on. Even though the epidemic was over, the triptych of viral evolution (selection, extinction, and resurrection) had never stopped playing out in the liminal spaces where humans and animals met: in live-animal markets, in bat-infested mine shafts, and in the places where humanity waged its assault on animals and their habitats. But SARS, Baric knew, had changed one thing: The days of thinking of coronaviruses as a laboratory specimen were over. Coronaviruses had all the weapons needed to become pathogenic, and the threat they posed was likely just beginning. Someone had to prepare humanity to meet whatever monstrous predator the *Coronaviridae* would produce next. The only problem was that neither he nor anybody else knew what it would look like. How do you prepare for battle against an enemy that doesn't yet exist?

WITHIN A COUPLE OF YEARS, SAVI WAS NO MORE. ITS WEBSITE STOPPED loading. Its accomplishments weren't recorded anywhere aside from in a handful of peer-reviewed studies. Its funding never went beyond two million dollars in seed money. And while the vaccine testing that Brunham and his team ran on animals served as the foundation for vaccine development by other scientists like Ralph Baric, beyond the halls of coronavirology, the initiative's scientific legacy was negligible. Most important, it never ended up bringing a viable SARS vaccine to market—not in a year, not in two years, not ever. The model of collaborative, nonprofit vaccine making on which Brunham had burned his scientific capital eventually failed. The defeat of SARS destroyed SAVI's raison d'être, forcing Brunham to confront a hard truth: nobody remembers the battles that were never fought.

SAVI's failure didn't just mean no SARS vaccine. It meant that there would be no template for a "plug-and-play" coronavirus vaccine waiting on the shelf in case the *Coronaviridae* family generated another pathogenic threat. And so, it was business as usual in the biotech market: academia would be where lofty scientific ideas budded, and pharma was where they would bloom and be harvested and sold.

In 2003, vaccine revenues made up only roughly 1.5 percent of all global pharmaceutical sales, making them among the least important health products on the market. It was a bitter irony, given that vaccines (especially those for polio, measles, diphtheria, and influenza) have arguably done more to improve human health than any other scientific discovery the twentieth century produced. And yet, vaccines, in the pharma world, were known as "low-margin commodities," meaning they didn't make much money for the companies that produced them. Even though tens of millions of children were vaccinated each year, and even though three million deaths every year from diphtheria, tetanus, whooping cough, and measles were prevented by vaccines, they didn't pad pharmaceutical giants' bottom lines, because the bulk of those vaccinations occurred in lower-income countries, where pharma charged little for such medicines. If routine vaccinations of millions couldn't make companies much money, it was a sure bet that vaccines for rare and emerging diseases were even bigger losing propositions. It's no wonder, then, that in the first two decades of the twenty-first century, not a single vaccine had been brought to market for novel pathogens like SARS, Ebola, or Zika, despite the sickness and deaths they cause.

2004, Shitou, Yunnan Province, China

Dr. Shi Zhengli had spent the last eight months of her life crawling around caves with absolutely nothing to show for it. When SARS-infected bats were found at the source of the first outbreak, a wildlife market in Guangdong, Shi was enlisted to probe what role the creatures had played, if any, in the emergence of the novel coronavirus. It was a task that would span years and earn her the nickname "the Bat Woman of Wuhan." It would also set her along the path that culminated in her ascension to the directorship of the Center for Emerging Infectious Diseases at the Wuhan Institute of Virology. Shi was driven by an intrepid fearlessness in hunting down and capturing bats, and her exploration of caves and rookeries ultimately represented one of the most consequential advancements in corona-virus science.

Shi was slim and wiry, and had the look of someone who would bend but not break. And though she sought to maintain a professional and steady composure, her face was highly expressive, her joy betrayed in a broad smile, and her frustration evident in her withering glare. She spoke in a surprisingly low and powerful voice, and when she became absorbed in her ideas, she would rock her head back and forth as if in prayer.

Throughout 2004, Shi, then a mid-career epidemiologist, visited limestone caves, dense forests, abandoned mines, and other potential bat hibernacula across Southern China, tracking and tagging the flying mammals as she went. Her sprawling exploration was driven by her hunch that bats played a part in the zoonotic leap that SARS made from animals to human beings. But after eight months of walking for miles over mountains, across jungles, and through underground caverns—crawling on her belly through narrow openings, avoiding stalagmites, catching bats in nets, performing anal and throat swabs on the captured creatures, and collecting their blood and feces—Shi and her team had nothing to show for their efforts. Not a single bat among the hundreds she had captured tested positive for any coronaviruses, let alone SARS. Despite the circumstantial evidence linking bats to the initial SARS outbreak, Shi began to doubt her hypothesis.

Shi had been using Ian Lipkin's PCR test approach to identify SARS virus in bats. That test, which detected viral particles in respiratory systems, was a highly effective tool for controlling the epidemic. But as a way to survey the extent of viral infection across a species, it had serious drawbacks. The major one was that PCR detected only active virus; given that SARS infection usually lasted a few short weeks, the chances of catching live virus in a single individual, bat or human, were slim. Much longer-lasting, though, were the antibodies that the immune system produced to counteract infection, which could linger for weeks, months, or even years. After eight months of fruitless searching, Shi realized that she had been going about it all wrong. If bats were, as she suspected, the natural reservoir for SARS, then it wasn't necessary to find evidence of active infection. Instead, finding remnants of immune memory circulating in their tiny bodies would be just as much of a smoking gun. So, Shi decided to ditch the PCR test and instead search for coronavirus antibodies using a broad-spectrum test that had long been a standard laboratory tool to detect the *Coronaviridae* family. When she retested the samples she had collected out in the field, it was as if a lightbulb—or, rather, a massive and unblinking spotlight—had been switched on. Among the hundreds of animals she had captured, probed, and sampled, coronavirus antibodies were everywhere.

Among one bat family—*Rhinolophidae*, tiny creatures known colloquially as horseshoe bats—fully a third of the creatures Shi sampled had antibodies for at least one coronavirus. In another group of bats, sampled in caves in Hubei, the province where the COVID-19 epidemic would later emerge in the city of Wuhan, almost three-quarters of horseshoe bats tested positive for coronavirus antibodies. Hubei was a glowing epicenter of bat coronavirus infection, but Shi found bats with signs of coronavirus infection in regions spanning the length of China. The high prevalence and wide geographic spread of infected hosts made it clear that bats were a natural reservoir for the *Coronaviridae*. This was further confirmed by the fact that the bats that had been infected didn't appear to have any lasting effects from their infections, meaning that the coronaviruses had found a highly efficient way to replicate in their bodies without doing much damage.

RALPH BARIC HAD BEEN PROBING FOR DECADES HOW EFFICIENTLY coronaviruses engineered themselves to jump between species. This in itself made them dangerous. Pair this adaptability with bats, one of the largest mammalian orders, and the potential for viral diversity became explosive. This made Shi's findings earth-shattering.

The revelation of the bat–coronavirus link was Ralph Baric's epidemiologic worst-case scenario. Bats make up a full 20 percent of all mammalian species, and they're found on every continent except Antarctica. Bats are also, like humans, deeply social animals: they roost together, individuals tightly packed one against the other in caves filled with fetid air, their breath comingling in the cold, dripping humidity and the dark. Left to its own devices, a coronavirus reservoir in bats wasn't in itself a problem, because the viruses would rarely have a chance to infect a human. But humanity's endless advances had brought our species closer and closer to bats' hibernacula and hunting grounds. It was this synanthropy (the merging of the world of humanity with that of wild animals) that caused an environmental shift, tilting the epidemic triangle toward a potentially explosive outbreak. With humans encroaching on their natural habitats, bats began silently roosting in obscure human places: in abandoned mines, under bridges, and in the rafters set high above market stalls where the trade in civet cats and pangolins, both living and dead, churned along each day. Baric pictured them shedding virus through their mouths and feces, infecting the wild animals below and the humans that handled them, the daily business giving ever-evolving coronavirus variants a new shot each day at making a successful leap into humans.

Baric did the math. There were, at last count, more than thirteen hundred known bat species around the world, including more than one hundred in China alone. Each bat species, based on the antibodies that Shi had discovered in the field, appeared to harbor about six to eight coronavirus strains. Bats being social creatures, the chances of infection were high; one bat colony discovered in a Texas cave numbered over twenty million. From an epidemiologic perspective, members of the same colony roosting together made up "pools of variance": roosting in close proximity, pool members would continually shed virus and reinfect one another over their entire lives, creating the perfect conditions for the wholesale swapping of genome sections. This genetic reassortment made the pools of variance the perfect nursery to create pre-epidemic coronaviruses poised for a jump.

With hundreds of millions of bats in the world, Baric calculated that there were tens of millions of potential coronaviruses that might at any moment make the zoonotic leap into human beings. It was, in Baric's words, a hidden bat virome: a network of viruses stretching across the earth's circumference. This virome wasn't a universe beyond our own, but rather a superimposition upon our world that only a rarefied class of scientific initiates could clearly see. It was a sobering if not terrifying realization: only one human coronavirus capable of creating a deadly epidemic had been identified, but it was a certainty that the future would invite many, countless, more.

The concerns Shi and Baric shared would soon be confirmed. A field study from the West African nation of Ghana—6,700 miles away from China—found that 10 percent of sampled bats contained SARS-related coronaviruses. When Ian Lipkin and his team sampled forty-two different bat species in Mexico, they turned up thirteen different coronavirus strains, twelve of which had never before been seen. In the United Kingdom, a sampling of Natterer's bats (two-inch creatures indigenous to the European continent) found that three-quarters of the tiny, leaf-nosed insectivores had evidence of infection. And still other bat coronaviruses, both active virus and antibodies, were found across Africa, the Levant, and the Indian subcontinent. The hidden bat virome was everywhere.

IN THE YEARS AFTER SARS, BARIC'S OBSESSION WITH CORONAVIRUSES led him to move into the realm of the truly arcane. The sheer breadth of the viral universe the *Coronaviridae* inhabited haunted him: tens of millions of individual strains, a veritable sea of microscopic pathogens dumbly probing every conceivable pathway to infect human hosts. The passage of time made it inevitable that the collision of different species and new variants would take place in an environment friendly to the pathogen's replication. Baric feared the next big jump and couldn't shake the worry that when it happened, humanity would be caught flat-footed. Luck had saved humanity from the poorly replicating SARS, but that wouldn't cut it the next time around. Baric needed to develop weapons to protect humanity against the next coronavirus that would emerge, despite the fact that he had no way to predict with certainty what it would look like. But he could make an educated guess. The *Coronaviridae* had traits that were shared across all its members. In human families, these could be a dark complexion or high cheekbones; among coronaviruses, they are nonstructural proteins like nsp12, which gives coronaviruses the power to copy themselves endlessly. Where the progeny were most likely to diverge, Baric knew, was in the spike protein, which mutated at three times the rate of other coronavirus components.

In 2005, Baric surveyed the tools at his disposal, gleaned over almost thirty years. It didn't take long. The sum total of scientific knowledge on human coronaviruses consisted of the identification of five distinct strains and a couple of promising SARS vaccine candidates, none of which had really gotten off the ground. It wasn't, frankly, much to go on, but it was a start. As a virologist, Baric could produce two basic weapons to counter viral pathogens: vaccines and antivirals. Vaccines train the immune system to destroy viruses as soon as they appear. Antivirals, though, artificially disrupt the monotonous replication of viruses, making clandestine attacks on the invaders when the body can't do so itself. Both can be incredibly effective: smallpox, a virus that caused at least five hundred million deaths over the past two centuries, was fully eradicated by a vaccine. While no vaccine for HIV exists, a cocktail of antiviral drugs, each of which targets a different stage of replication, has fully curtailed HIV's capacity to kill (though getting the medicines to the thirty-eight million HIV-positive people around the world is an ongoing battle). The catch, though, is that both vaccines and antivirals have only ever been developed against known viruses, and even then, the road to creating them was long and difficult, and often ended in failure. It's one thing to battle a known pathogen, but when the virus you're trying to target doesn't exist yet, the challenge becomes almost absurd. Baric tried anyway.

Even though the future coronavirus pathogen was unknown, Baric figured it wouldn't be entirely unfamiliar. Chances were that it would look something like the SARS viruses he had collected from the three stages of the epidemic, which included those strains that naturally resided in bats but that had proven hardy enough to survive in multiple other species, including humans.

Baric had collected a menagerie of coronaviruses (not only SARS strains, but also MHV, NL63, OC43, and many others), which he manipulated freely in his lab. Since his time as a postdoc in Los Angeles, he had also become adept at recombining parts of coronavirus genomes, which had become a common practice among virologists as technology improved throughout the 1990s. All these pieces—the various coronaviruses he had on hand, his skills in synthetic recombination, and his understanding of the minutiae of the structural makeup of the family—led him to a momentous decision. If he really wanted to know what the future looked like, it would be simple enough in practice to alter the viruses he had in order to anticipate the strains that might yet come. The trick was figuring out what vulnerabilities that future pathogen would exploit to infect our species.

Baric began to dream about that future threat. Until SARS, his lab had focused on prodding and probing existing viruses to see how they would evolve under pressure, as with his experiment that transformed a mouse coronavirus into a hamster-infecting superpredator, a class of experiments broadly known as gain-of-function research. But in 2004, he slowly began to nurture a menagerie of synthetic viral abominations specifically designed to help him anticipate the most likely pathogen the *Coronaviridae* might yet produce.

Baric had been for years developing a technique called cDNA cloning, which amounted to reverse-engineering full viral genomes from the impermanent scraps they left behind in the process of copying themselves. When viral RNA replicates, it produces complementary DNA (or cDNA) strands that help the virus take over the functions of a host cell. As a bonus, cDNA strands also match up perfectly, like the cast of a statue, against the genome of the virus, meaning that they could be artificially stitched together and then copied in negative to re-create an original strand of viral RNA. Baric was one of the early adopters of cDNA technology and had used it to rapidly clone SARS in his lab using the genome sequence that Brunham and Marra had mapped; his feat pointed the way for researchers who wanted to run experiments on the virus but didn't have a reliable supply. For Baric, who published more than four hundred peer-reviewed articles picking apart different facets of coronavirus replication, solving the supply problem was only the first order of business. After SARS, he also recognized that cDNA techniques had the potential to go beyond simply re-creating viruses that already existed. If each cDNA strand was made up of a specific region of a coronavirus genome, then why couldn't he simply add different cDNA strands together to make a whole new strain? And if he could, why wouldn't he? After all, the point was to predict the future, and you don't do that by reverse-engineering the viruses that someone has already mapped.

Baric's first attempts to create new synthetic viruses hewed closely to known isolates. After cloning SARS using cDNA components in 2003 from the Urbani strain (named after the late Dr. Carlo Urbani, the WHO's Vietnam country director who died after getting infected), Baric used cDNA to reverse-engineer the multiple strains of SARS that resembled, as closely as possible, those versions of the virus that predominated in the epidemic's first, second, and third waves. He started with fifty-four different SARS sequences taken from patients at different points in the epidemic. He then used cDNA techniques to create five synthetic variants; each variant had a spike protein that differed in its ability to hook on to and fuse with human cell walls, mimicking the evolution of the SARS virus during the epidemic from a stuttering, inefficient transmitter to an elegant predator. It worked like a charm. In mouse models designed to mimic the virus's impact on humans, his artificial "first-phase" variants couldn't replicate at all, and the variants built to mirror the second and third waves of the epidemic replicated more and more efficiently. What's more, while young mice infected with the later-phase variants were free of disease, the old mice, once infected, developed classic SARS-like symptoms and eventually died. Baric had told the tale of an entire epidemic in miniature.

The point of the experiment wasn't just to relive the glory days of SARS. It was to test vaccine candidates on a panel of coronaviruses that mimicked the strains that had come before and that might come again. Baric knew that vaccines that didn't accurately match their targets could weaken human resistance and inadvertently make people who were inoculated sicker. It was an issue that had preoccupied him for years. "If SARS kind of went extinct in humans but it was still present in animals," he explained, "we want to know which vaccines would work against the strains in animals. Because that's what's going to come out next: it's going to be another introduction event. And we know there's always introduction events." In a follow-up experiment undertaken over five years, Baric and his team tested SARS vaccine candidates in mice and found that many were exceptionally effective against the specific strain of the virus they were designed to attack. The problem, though, was that when the vaccines were tested in mice that had been infected with slightly mutated SARS variants (variants that Baric had artificially created in his lab), the vaccines caused the animals to become more severely ill than the mice that weren't vaccinated. Among older mice (known as retired breeders), the negative reaction was even severer. In short, the SARS vaccines were more likely than not to make future coronavirus infections deadlier, even if the virus strains were closely related to coronaviruses that had come before.

Baric was one of the few scientists who saw the stakes clearly. It wasn't that an eventual SARS vaccine might work or not. It was that a vaccine could be either an antidote or a poison for a future pandemic-ready coronavirus, and there might be no way to tell them apart until it was too late. Baric had spent decades prior to SARS obsessed with the abstract puzzle of the *Coronaviridae*. When his obsession turned up in the real world in the form of SARS and began to kill people, he couldn't shake the feeling that it was just the beginning of the horror. As the years went by, he hardly felt their passing, so deep had he fallen into viral time, watching mutations emerge under the fluorescent glare of laboratory lights and wondering which of them heralded the perfect future threat to come.

INTERMEZZO—THE TONGGUAN MINE

August 2012, Mojiang Prefecture, Yunnan Province, China

The derelict mine lay hidden among the lush tropical vegetation on the mountains of Yunnan Province, sixty miles from China's southern border with Laos, Vietnam, and Myanmar. Though it had once been used to tap a copper vein, the central shaft leading into the rocky earth had fallen into disrepair after its long abandonment. In the absence of human industry, the mine had become a haven for vermin (rats, musk shrews, and bats) attracted by the promise of a cool, damp hibernaculum. In 2012, six miners were hired to clean slag and remove bat feces from the mine shaft in preparation for its renewed operation. After four days, two of them fell ill with pneumonia; ten days later, the other four caught the infection. All shared similar symptoms: difficulty breathing, dry cough, fever, aching limbs, and headache. As their conditions worsened, the miners were hospitalized and subjected to a battery of tests. Experts from across China—including those who had first won fame beating back the SARS epidemic—were patched in to provide remote consultation on the curious cases. But the miners' symptoms worsened. Four developed acute respiratory distress syndrome and were put on mechanical ventilators. Some of them began spitting up blood. Eventually, three of the miners were released from the hospital. Within one hundred days, though, the other three had died. The cause of death was, officially, unknown.

News of the outbreak quickly spread through official channels, eventually reaching Dr. Shi Zhengli, by then the director of the Center for Emerging Infectious Diseases at the Wuhan Institute of Virology. The message was clear: find the source of the disease. Nobody needed to explain the stakes. SARS had emerged only a decade earlier and had demonstrated the full scale of the tragedy that could arise from clusters of new, unexplained pneumonias. Failing to identify novel viruses or to quarantine areas could very quickly escalate into a worldwide panic. So, along with a team of researchers and scientists, Zhengli made the 1,100-mile journey from Wuhan, in China's industrialized north, to southern Yunnan Province, where the Tongguan Mine opened like a clotted vessel into the earth.

The cave stank like hell, said Shi, the odor a fetid mix of bat guano, rat droppings, and a fungus feasting on the fecal matter, made all the worse by the humidity. No matter: Shi had arguably more experience than any other virologist in cave spelunking, and the odor was just a part of what she had signed up for. She also knew that there was only one way to complete the mission she had been given, which was to create a viral scan of the mine shaft. So, she and her team got to work, fanning out across the shaft and its surrounding areas to trap bats. The creatures were everywhere. The mine shaft had been settled by six different species of bats, all of which had set up overlapping colonies in the chaotic, humid gloom. Shi's team combed the area, spotting and identifying individual bats and collecting fecal samples. Ultimately, over a period of two years, her team returned four times to the Tongguan Mine, sampling 276 bats in the search for the origins of the illness that had felled the miners.

When all was said and done, half of the almost three hundred bats tested positive for coronaviruses. Worryingly, all six bat species also harbored multiple strains—152 in all—opening up the strong possibility that the bats in the mine shaft were a particularly dangerous reservoir for viruses engaging in genetic reassortment, the most effective way to produce new coronavirus mutants. Many of the strains that Shi identified were closely related to existing bat coronaviruses and fell into the *Alphacoronavirus* genus (which includes pig, bat, and other mammalian strains, as well as NL63 and 229E, two benign human CoVs). Two of the strains, though, were novel betacoronaviruses, the *Coronaviridae* lineage that had, before the emergence of SARS, been nothing more than theoretical. Of those two novel betacoronavirus strains Shi's team identified, one was closely related to a bat coronavirus sampled in Ghana. The other, later dubbed RaTG13, was a relative of SARS, but thankfully a distant one.

In the publication stemming from Shi's time combing the Tongguan Mine for fecal pellets and trapping bats, Shi and her colleagues pointed out that RaTG13, named for the bat species (*Rhinolophus affinis*, a horseshoe bat common to China), place (Tongguan Mine), and year (2013) in which it was discovered, was only 77 percent identical to SARS, making it more distantly related to the human pathogen than many SARS-like bat coronaviruses that had already been discovered. Taken on its own, this finding brought a sigh of relief in observers worried about another SARS outbreak. The closer a new coronavirus strain was to SARS, everyone assumed, the more likely it would be primed for human transmission. This made RaTG13 a dead end in an investigation of what had killed the miners.

But there was more. The intersection of these novel strains in the abandoned mine shaft, and Shi's presence there, was no accident: the Chinese government's distress over the possibility of another SARS outbreak had made it almost inevitable. Where viruses, bats, and pneumonias converged, so, too, did Shi and her team. Strangely, though, while the story of six miners falling ill from pneumonia after coming into contact with bats should have raised international alarm, it caused barely a ripple of interest. This wasn't merely by chance: the lack of attention was helped along by reporting from within China that implied that the deaths of the miners had been caused not by a SARS-like coronavirus but by a paramyxovirus (from an unrelated viral family known mostly for sickening horses), which Shi's team had also identified in some of the countless rats scurrying through the mine's warren-like structure. In a story from *Science* magazine titled "A New Killer Virus in China?," published shortly after Shi's findings were released in 2014, the new paramyxovirus was described as "more of a curiosity" than a real threat. Oddly, no mention was made in the article of RaTG13 or of any of the hundreds of other novel coronavirus strains that Shi had discovered in the cave. "The three victims in Yunnan succumbed long before scientists arrived on the scene," the article concluded, hinting that the real pathogen would likely never be known. The investigation, the reporting implied, was an epidemiologic dead end, one of those strange unexplained phenomena that are forgotten by the next news cycle. There was and would be nothing more to learn.

And yet, within China the story did not stop there. In May 2013, an industrious master's student at Kunming Medical University in Yunnan Province published a thesis with the provocative title "The Analysis of Six Patients with Severe Pneumonia Caused by Unknown Viruses." The thesis, written in Chinese, analyzed all the data surrounding the illness and deaths of the miners, including lung X-rays, blood samples, viral work-ups, and the course of the illnesses that felled the miners. After retesting some samples in the zoology lab at his university and reviewing the entirety of the data that Shi Zhengli had collected, the student was adamant that he had pinpointed the cause of the miners' illness. "Based on the above-mentioned cases and related searches," the thesis concludes, "the unknown virus that led to the severe pneumonias could be: A SARS-like CoV from the Chinese horseshoe bat, or a SARS-like CoV from another bat." Either way, a coronavirus was responsible.

It was an audacious scientific statement and exactly the kind of galling and risky work that junior scientists were meant to undertake. And perhaps if it had been written in English, speculation may have bubbled up among those keen observers of emergent coronaviruses that the deaths of the miners were, in fact, caused by a relative of SARS that had emerged from the depths of the hidden bat virome. Instead, the mine shaft was shuttered, and the story faded into obscurity.

And perhaps, even then, it wouldn't have ended there. But in a strange case of timing that seemed to prove Baric's worst fears, a new threat emerged in the midst of Shi's investigation. This new pathogen, though, was nowhere near her or the hibernacula that had become her second home. The jump had taken place half a world away.

Part II

Chapter 6

SOME CAMELS ARE MORE EXPENSIVE THAN HUMANS

June 12, 2012, Bisha, Saudi Arabia

T HE TEMPERATURE HAD BEEN HOVERING AROUND 104 DEGREES
Fahrenheit for weeks. This was good news for the businessman's date palm orchard, but the heat and humidity made him feel
wretched when the first symptoms emerged in the middle of June,
in the midst of the hottest weeks of the year. At sixty, he had done
well in the kingdom, or well enough, having attained something of
a model life: a thriving business, some land, four pet camels, and
three wives. The large paint warehouse he owned made good money,
enough to provide separate houses and generous allowances for his
wives, and he had even saved sufficiently to break ground on a house
that would all but seal the deal with his soon-to-be fourth bride. His
finances had been borne aloft on Bisha's building boom, which had
risen along with Saudi Arabia's increasing prosperity. Since the
opening of the King Fajad Dam in 1998, agricultural stability had
arrived in the region, and the spring floods, while still severe, were
largely contained. The economy had been further buoyed by the dig-
ging of new oil wells in the kingdom's southwest, and it was a time
of plenty. There were smaller joys, too: the paint warehouse's large
garden, full of fruit trees and buzzing insects, imparted a modicum
of elegance to the commercial site, especially when the desert sun

ebbed across the Arabian Peninsula and the roar of cars outside was replaced by the quiet sound of flapping wings and twittering echo-location.

But the next day, the businessman's symptoms got worse: fever, cough, fluid-filled lungs, and shortness of breath. The man wasn't the picture of health. He was obese, though this was offset by a lack of heart conditions or any other major illness, and if at sixty he wasn't exactly young, in the kingdom he was still considered in his prime. Immediately, the man traveled to the city of Jeddah, three hundred miles northwest across the Arabian Desert, where he was admitted to the Soliman Fakeeh Hospital, an imposing glass-and-concrete private hospital that boasted a modest virology lab, for diagnosis and treatment. But despite the state-of-the-art treatment he received, the fever never subsided, the heat of his body eventually matching the extreme temperature of the desert sun, his intubation in the hospital's modest ICU accomplishing nothing. The pneumonia advanced, his kidneys began to fail, and his body became unable to fight against the relentless advance of the illness.

On June 24, 2012, the businessman from Bisha died. In life, he was a person like any other. In death, he was transformed into the index patient for a terrifying new epidemic, which signaled that the age of pathogenic human coronaviruses had only just begun.

FOR ALL ITS IMPOSING GRANDEUR, THE SOLIMAN FAKEEH HOSPITAL was no scientific hub. Indeed, its small virology lab was something of an afterthought. The lab wouldn't even have existed were it not for three empty rooms on the hospital's sixth floor and the resolve of one person, Dr. Ali Zaki, an Egyptian-born virologist first recruited to Saudi Arabia in 1993. Over two decades, Zaki, a short man with a slightly mischievous grin and gentle almond-shaped eyes, had turned those three rooms into a spartan but functional cell culture lab, antibody testing lab, and molecular biology lab, all of which were at the ready when the index patient was admitted on June 13,

2012. Unbeknownst to Zaki, the death of the man from Bisha would set in motion the end of his own life in Saudi Arabia.

As the man from Bisha lay dying in the ICU, Zaki was given samples of his blood and saliva and asked to run routine diagnostics to isolate whatever respiratory pathogen was causing him to die. All the standard tests—for influenza A and B, for adenoviruses, for HIV—came back negative. So, Zaki reached further into the arcane, testing for viruses that, while unlikely, might be the culprit—viruses like Nipah and Hendra, both of which kill most of those they infect. Finally, at a loss, he used a pancoronavirus test (which detects all known coronaviruses), which yielded a positive match. He was stricken by the results. It had been a decade since SARS, which was considered by everyone to have been fully eradicated. And yet, Zaki thought, perhaps the virus could still have been lingering in some animal reservoir. He assumed that he had stumbled upon the first signal of an imminent second wave, ten years too late. But when he was finally able to get his hands on SARS-specific tests, the match was negative. It was a pathogenic coronavirus, to be sure, but a new one.

Unable to sequence the novel viral genome in his lab, Zaki nevertheless wanted an answer. He also knew that the Saudi authorities wouldn't approve a request to send a dangerous and unknown virus abroad for testing, but the potential danger to humanity if he didn't was far too great. So he had a sample sent to Dr. Ron Fouchier, an expert in emerging viruses at Erasmus University in Rotterdam, Netherlands. It was no sure bet that the sample would arrive intact: Zaki's lab wasn't equipped with cold storage, a prerequisite for maintaining the integrity of live viruses in transport. So, he improvised, placing the sample at room temperature in an African green monkey tissue culture, which he put in a tube and sent in a biohazard container to the Netherlands. It was a leap of faith, one predicated on the virus's capacity for predation: if it could consume the monkey tissue and keep itself replicating long enough, then humanity would have a shot at uncovering its identity.

The package was loaded onto a plane leaving Saudi Arabia, crossing the Levant and the bulk of the European continent to arrive, finally, in Amsterdam, where it was then transported to Rotterdam and to the laboratory of Ron Fouchier at Erasmus. Upon the sample's arrival, the Dutch virologist carefully opened the package and examined its contents. Against all odds, the pathogen was still replicating, saved by its rapacious hunger. Without wasting any time, Fouchier sequenced the sample and revealed, for the second time in just a decade, the architecture of a new and deadly coronavirus. The man from Bisha would not be the last of its victims, and it would not be the last of its kind.

There was a chance that the Saudi Arabian government might be able to control the virus on its own. Regardless of how much of a threat it presented, though, one thing was clear: having the world's best virologists on the case could only help prevent a catastrophe. And this left Dr. Ali Zaki facing his second hard choice: he could keep quiet and notify only the Saudi authorities about what had killed the Bisha businessman, or he could alert the world to the new pathogen, which would mean flying in the face of in-country protocols that punished publicly sharing sensitive public health data with years of jail time or even, potentially, execution. It was, in the notoriously strict kingdom, a serious transgression at the best of times. In 2012, though, perceived threats to the monarchy's sovereignty were even more loaded: the nascent Arab Spring protests, which had begun in December 2010, were roiling politics in the region and placing leaders on edge. This had made the leaders of Saudi Arabia, already paranoid over internal threats to their control, even more sensitive to perceived sources of chaos. The then ruler King Abdullah bin Abdulaziz Al Saud had also shown himself willing to use the full power of the state-controlled media to marginalize dissent and arrest and even execute those who proffered criticism of his rule. Zaki assumed that the best scenario was that he would lose his job. The worst was too horrible to consider. Still, that paled in comparison to what might happen to the world if he remained silent.

On September 20, 2012, at 3:51 P.M. Eastern Standard Time, an

alert popped up on ProMED, the global Listserv run by the
Massachusetts-based International Society for Infectious Diseases.
"Novel Coronavirus—Saudi Arabia: Human Isolate," read the subject line. Ali Zaki had made his move. The short note outlined the
basic case of the businessman from Bisha, including his untimely
death and Zaki's decision to send a sample of the virus out of the
country and into the Netherlands for genetic sequencing.

As soon as the post appeared, the full wrath of the Saudi Ministry of Health rained down on Zaki. The ministry accused him of
criminally transporting the sample of what was now being called
Middle East respiratory syndrome, or MERS, out of the country.
Zaki knew that it was only the start of his troubles. So, five days
later, he boarded a flight to Cairo, leaving all his personal belongings
at his home in Jeddah, never to return.

After forcing Zaki to flee the country he had called home for two
decades, Saudi authorities quickly took steps to scrub his legacy. The
lab he had built at the Soliman Fakeeh Hospital was dismantled, and
his MERS samples were subject to extreme heat and destroyed. If
not for the sample Zaki had sent to Fouchier, no record of the virus
culled from the index patient would have existed. The government
then began spreading the rumor that Zaki was far from the hero he
had made himself out to be. Referring to him obliquely as "the Egyptian doctor," a senior Saudi health authority claimed that Zaki's lab
at the Soliman Fakeeh Hospital had numerous regulatory violations
that compromised its biosecurity. What's more, Zaki's decision to
send the Bisha sample to the Netherlands hadn't been heroic but
merely a naked attempt at personal gain. Health authorities then
circulated a rumor that it was Zaki himself who had, instead of trying to alert the world to the new coronavirus, inadvertently unleashed
it, killing the index patient in the process. The doctor, they claimed,
had "either intentionally or inadvertently not followed national established procedures" and was responsible, therefore, for spreading
MERS. This intersection of politics and infectious disease, it was
clear, was as much a defining feature of emergent pathogenic coronaviruses as their mutable spike proteins, and just as deadly.

WHEN WORD SPREAD THAT A NOVEL CORONAVIRUS HAD BEEN ISOLATED from a patient in Saudi Arabia and was spreading across the region, Malik Peiris was keen to help. He was, after all, the first to isolate SARS and had been spending decades honing his skills in predicting when new respiratory pathogens might emerge. Peiris was among a handful of scientists in the world with the requisite combination of virological knowledge (like the environments necessary for viruses to jump from animal to human populations) and on-the-ground investigative skills (like what kinds of questions to ask of infected people) to prevent epidemics from spreading. With his courteous and empathetic manner, Peiris also had a knack for making friends across the world—never a bad thing when novel epidemics can lead one to strange and unexpected destinations. Before he could help, though, he needed the virus, and that was no sure thing.

Information in Saudi Arabia was tightly controlled at the best of times, which meant that getting a viral sample out of the kingdom was next to impossible unless you were willing to break the rules. But Peiris, who always exuded a certain tenderness, had friends all over the world. One of them was Ron Fouchier. After having sequenced the virus, Fouchier didn't hold tight to his new discovery. Knowing that Peiris, who had an unparalleled reputation for epidemic prevention, was on the hunt for the source, Fouchier was more than happy to share. With a deadly new pathogen emerging, it was no time for reflexive academic protectionism.

After the SARS epidemic, Peiris had barely caught his breath before wrestling with the 2005 H5N1 avian flu pandemic, which killed hundreds of people, required the culling of 1.2 million poultry, and led to the death of 140 million birds. It was a viral threat that never quite receded entirely, with occasional outbreaks popping up spontaneously across the global supply chain of farmed and butchered bird meat. This had made Peiris instinctively eager to trace zoonotic jumps as quickly as possible when new viruses emerged. The source of the infection had to be capped like the sealing of a poisoned well. With the viral sample from Fouchier in hand, Peiris set about following his epidemic-prevention playbook: he rapidly developed an antibody test with the intention of sampling animals to find which one was the reservoir for MERS. While SARS had emerged from errant interactions among bats, other wild animals, and human beings, avian flu was the product of repeated and predictable exposures between humanity and the millions of livestock our species breeds for food. Though he was dealing with a coronavirus, many of which were found in bats, Peiris's hypothesis was that the animal reservoir for the new pathogen would most likely be livestock or pets, just like with avian flu. It was a hypothesis, nothing more.

Before he could test it, though, he ran into another problem: he needed to test animals from inside the kingdom. But the political response to the MERS outbreak in Saudi Arabia was deteriorating, and public health officials in the country weren't willing to work with Peiris and others to get a handle on the crisis. Yet viruses replicate and spread, and the possibility existed that MERS had been spreading asymptomatically across the Saudi population and beyond, mirroring other deadly respiratory infections like avian flu. But without a reliable test for the virus, there was no way for Peiris to know how far the novel coronavirus had spread, how many victims it had culled, and how terrified the world should be about it.

Every single day of delay was another twenty-four hours during which the virus was given a reprieve. Peiris had battled avian flu, which had a case fatality rate of 60 percent. He had isolated SARS, which killed between 10 and 15 percent of those it infected. MERS, from what he had gleaned, was far deadlier, portending an ominous new phase in the history of the *Coronaviridae*. Peiris knew the playbook for preventing viral respiratory epidemics by heart; he had literally written whole chapters of it. His *Severe Acute Respiratory Syndrome: A Clinical Guide*, first published in 2005, recounted in minute detail each step that was needed to stop pathogenic coronaviruses from gaining a foothold in human populations. Public health officials in Saudi Arabia weren't blind to

Peiris's expertise and reputation, and when he pushed to enter the kingdom, they engaged in polite negotiations. But Peiris soon realized that it was all a front: the Saudi government had no intention of letting him into the country to test its livestock; they were just hiding their intransigence behind a veil of civility. In his exceedingly reasonable way, Peiris encouraged them to relent, but reason had long since left the conversation. He and his vast stores of knowledge were left idle. And still, the virus continued to kill.

Peiris waited obsessively for reports of new MERS cases, and his heart sank when each new victim was added to the death toll. Desperate to get into the kingdom to begin testing animals for MERS, he found his path blocked. When he sought approval from the Saudi government to use his MERS antibody test on domesticated animals there, the official response was that his test, being so new, required scientific validation before the authorities would sanction it. To Peiris, this smacked of duplicity. It was, admittedly, more complicated to design a test that detected antibodies rather than viruses, but testing directly for virus was also, as Shi Zhengli had discovered, a supremely ineffectual way to gather epidemiologic data on how widespread a virus might be in an animal reservoir. It was hard enough to find evidence of infection with a specific virus in a specific species in a specific geographic area; having also to find the animal hosts at the exact moment they were infected considerably narrowed the chances of finding the host reservoir that had spawned the virus.

The public line among Saudi government officials was that MERS was a domestic problem that could be contained without international interference. The only way their minds would be changed was if the virus, that unblinking automaton, proved them wrong. By the laws governing epidemic spread, it shouldn't have. MERS was slow-moving, its overall reproductive ratio (also known as R_0), or the number of additional people who became infected from each initial case, was below 1, meaning that the epidemic should have been contracting, not getting larger. But that number varied wildly depending on the context, with some outbreaks in

Saudi Arabia reporting an R_0 as high as 6, meaning that the deadly virus was spreading voraciously. On the ground, this meant that what should have been a domestic Saudi problem that could be easily tamped down had become a tenacious, growing crisis. Patients would present at Saudi hospitals with lungs so damaged that they had become rigid and unable to absorb oxygen; days later, the patients' kidneys would fill with cytokines and fail, leaving them dying of a cruel combination of hypoxia (lack of oxygen) and renal failure. The healthcare workers caring for them would invariably get sick as well, with as many as half of them dying despite adhering to the rules that protected hospital staff during the SARS epidemic: proper personal protective equipment (PPE), quarantining, and chemical sanitation of any surface that might harbor virus. Still, though MERS cut an unrelenting path through its human victims, these deaths were considered a domestic Saudi problem.

But then a forty-nine-year-old Qatari man checked himself into a hospital in respiratory distress and a rapidly deteriorating condition. With the hospital unable to care for him, he was flown directly to London, where, in an isolated, pressurized room, a team of specialists in full PPE worked to save his life. They failed: though they were able to stabilize his breathing, his kidneys stopped working, and he quickly died. Then an American healthcare worker visiting Saudi Arabia fell ill. The worker's travel path back to the United States was an epidemiologist's worst nightmare, involving a flight from Riyadh to Heathrow, with a short stopover, a change of planes, and then another flight, to Chicago O'Hare. From there, the worker took a bus to a city in Indiana. Soon after, complaining of respiratory symptoms, fever, cough, and shortness of breath, he presented in an emergency department at a community hospital in the city of Munster, where he tested positive for MERS. From then on, details were scant; the CDC, fearing all-out panic, declined to provide any more specifics, other than to assure the world that the risk of infection was low. This was just the start: there was a cluster of cases in Riyadh and then outbreaks in the United Arab Emirates, Jordan, Algeria, Germany, Iran, and halfway across the globe, in South

Korea. MERS, a slow-moving and deadly predator, was tenaciously making its way across the world.

Within six months of Zaki's first alert about the novel pathogen, thirteen people—all of whom had some connection to the Middle East—became infected with MERS, seven of whom had died, giving the coronavirus a shockingly high case fatality rate of 54 percent. For those infected with MERS, it seemed, survival was a coin toss. Peiris had continued to plead with the Saudi Arabian government to let him begin testing so that the source of the infection could be identified and more deaths averted—to no avail. Alternating between hubris and denial, Saudi officials refused to listen to reason. For Peiris, it was like knocking his head against a brick wall. There was just no way he was going to break down the barrier keeping him out of the autocratic society. So, with lives on the line, he decided to switch tack.

October 2012, Saudi Arabia

While Malik Peiris had been trying to barge his way into Saudi Arabia, Ian Lipkin slipped in through the back door. Just a few weeks after the Bisha index patient died, and with the entire global community in the dark about the threat MERS posed, Lipkin received an invitation to enter the kingdom—making him an *n* of 1 among the phalanx of international scientists seeking access to study the novel virus. Lipkin had an unparalleled scientific reputation that had been recently burnished by a consulting gig he had done on the blockbuster pandemic film *Contagion*. In the tight circles of government-funded public health, though, he was better known for something equally as important as scientific expertise: discretion.

Lipkin's trip to Beijing in 2003 during SARS was a case in point. He had come with ten thousand highly sensitive SARS tests, had schooled the local authorities on effective public health protocols, and had left without throwing anyone under the bus or pointing the finger at government blunders. It was, to some, an antithetical approach to epidemic prevention. Visiting a hot zone only to kowtow to local leadership meant that you were sacrificing lives to the epidemic, when the point of science was to be a shining beacon of truth that cut through the morass of government corruption, misinformation, and stonewalling that so often accompanied the emergence of new pathogenic threats. Lipkin, of course, saw things differently. His success was the product of a deep empathic sense, developed at an early age and honed by training, that gave him access to otherwise inaccessible places. "I read body language and eat things that other people won't eat," he said, "and breach these cultural barriers that other people just won't." There was something else, though, a cardinal rule Lipkin had maintained for his scientific forays: he would never share details of his trips nor publicly chastise local governments, no matter how autocratic or ineffectual. This didn't stop him from privately sharing sensitive information with Dr. Anthony Fauci, the director of the U.S. National Institute of Allergy and Infectious Diseases, officials at the World Health Organization, and contacts at the U.S. Centers for Disease Control and Prevention, all of whom were deeply involved in the global response to emerging epidemics. It meant, however, that he shared his thoughts on background, always without attribution.

The epidemic triangle describes the shifting relationship among pathogen, host, and the environment that causes epidemics to arise. Lipkin's gambit was that the environment was inextricably connected to government. If epidemiologists were serious about controlling the forces driving epidemics, then they would fail if they weren't able to manipulate political forces as expertly as they could develop diagnostic tests. It may have been unpleasant or ethically complex to deal with governments, especially authoritarian ones, without lapsing into public criticism, but for Lipkin, this was tantamount to shirking his duty as an epidemiologist. Ironically, his willingness to refrain from publicly criticizing governments had given him a reputation as a free agent beholden to no one, which in turn had allowed him to assiduously build political connections around the world. Others could quibble with his approach; the fact was that his complicated commitment to self-censorship was enough to grant him access to where he needed to be. This had been the case with the SARS epidemic, during which the Chinese Communist Party cajoled him into flying to Beijing. And so it was with the MERS epidemic, when the government of King Abdullah bin Abdulaziz Al Saud, "Custodian of the Two Holy Mosques," requested his assistance in relieving the kingdom of its pathogenic scourge.

Lipkin was keen on tracking the novel coronavirus, and he fully expected resistance from the Saudi government—he had dealt with authoritarian governments before, and this was part of the job—but once he landed in Riyadh, he realized that the country's rulers were going to be far more obstructionist than he had anticipated. Along with Shi Zhengli, Lipkin had been part of the team that had identified bats as a natural reservoir for multiple coronaviruses, including SARS and the separate viral sublineage (Group B betacoronaviruses) to which MERS belonged. He believed that the most likely scenario was that MERS was transmitted from bats directly to humans or that it emerged through an intermediate host animal. But when he explained that tracking MERS to its animal source would naturally require taking samples from bats, the Saudi health officials he was dealing with offered a bizarre response: bats, he was told, did not exist in Saudi Arabia. Lipkin, frustrated at the intransigence, was stuck having to deal with a problem that seemed metaphysically stupid, not to mention a massive waste of time. So, before he could even begin his epidemiologic investigation in earnest, he and his team were forced to fan out across the country, taking photos of bats, setting up nets to trap the flying mammals, and sending the evidence back to the Saudi Ministry of Public Health. It was absurd, but he was in no position to dictate the rules. Eventually, his local interlocutors acquiesced. It was a totally unnecessary win that presaged just how long and arduous the road to tracking MERS would be. And it further bolstered Lipkin's belief that although viruses moved in predictable ways, governments would always remain capricious actors in the face of epidemics.

Once he proved that Saudi Arabia was indeed home to bats, Lipkin set out to track the viruses they harbored. From Riyadh, he took the ninety-minute flight south to Bisha, landing in the diminutive agricultural town to trace how the environment, that shifting point of the epidemic triangle, had caused the paths of the index patient and the new coronavirus to fatally meet. It was time to walk in the steps of the lost Bisha businessman and, like a ghost forced to haunt the places at the heart of its turmoil, take from the living to understand the dead.

Lipkin's team began by interviewing the man's family and employees for any clues about possible transmission routes. They quickly struck out. Not a single person they talked to remembered seeing a bat anywhere near where the man lived or worked. Undeterred, the team followed the paths of the businessman's life, searching his paint warehouse, the small enclosed garden the warehouse fronted, and the date palm orchard beyond, its seventy-odd trees weighed down by the sweet and pulpy fruit that ripened over the summer months, attracting insects and those creatures that preyed upon them. Lipkin's team staked out the garden and the orchard, along with abandoned wells and ruins within an eight-mile radius of the index patient's home, in search of bats, while two other teams hunted for the winged creatures near Riyadh, Saudi Arabia's capital city, and Unaizah, in the kingdom's north.

They were not disappointed. Using mist nets and harp traps (the preferred tools of bat hunters), made of thin fibers that confound bat echolocation, the team managed to capture ninety-six bats spanning seven different species in just three weeks. After subjecting the bats to tissue sampling by punching small holes in their wings, collecting their blood, swabbing their throats, and taking rectal and fecal samples, the team placed the vital specimens—over one thousand in all—in lysis buffer, an agent that breaks down cellular material to allow molecular biologists to probe the strange viral invaders that lurk within cell walls.

Lipkin had early on made arrangements to ship all the samples collected in the kingdom back to his lab at Columbia University in New York City. His team did their part, dutifully storing the bits of wing tissue, fecal matter, blood, and saliva in liquid nitrogen, which kept them at a frigid minus 112 degrees Fahrenheit, sufficient to preserve any genetic material from the viruses they might be harboring. Blanketed in cold, the samples were sent from Riyadh to the United States, the home stretch of a journey that would end with an analysis of the viral genomes, followed by—what Lipkin assumed would be—unequivocal proof that bats were a natural reservoir for MERS.

But mere miles from their final destination at Lipkin's lab, disaster struck. Zealous U.S. Homeland Security officers at airport customs insisted on poking around inside the suspicious-looking package, despite its biohazard labels and warnings that doing so might destroy its contents. These warnings unheeded, the officials broke open the cold storage container, causing the liquid nitrogen to evaporate within minutes in the featureless customs back room, exposing the vials to room temperature air. Lipkin, desperate to salvage what he could, pleaded with the authorities to return the samples. The Homeland Security officials remained unmoved, despite his explanation that they contained the secret to controlling a novel pathogen about which we knew very little and that was killing most of those it infected. Lipkin waited, increasingly despondent, as the minutes ticked by. The weeks of work he and his team had done to trace the life of the index patient and capture the animals that likely harbored the virus, all of it encapsulated in the tepid vials, was disintegrating before his very eyes. He likened the samples to fingerprints at a crime scene: the best way to ID the murderer was a fingerprint match from all five fingers on a murder weapon. As the hours slowly passed, he pictured the fingerprints fading one by one, leaving something far less than incontrovertible proof. It was agonizing.

Forty-eight hours after they were confiscated, Lipkin's viral samples were finally returned to his lab, thawed, damaged, and barely intact. Furious but undaunted, Lipkin forged ahead, hoping against all odds that some of his work could be salvaged. The samples were immediately purified, and then his team tried desperately to extract the genetic information that might remain in the strands of viral

RNA. It was a disaster. RNA, notoriously fragile even under optimal lab conditions, was no match for two days on a shelf in a customs back room.

Lipkin's team tried to amplify the samples one by one, using PCR tests matched to parts of coronavirus genomes that were conserved across the family. (PCR, or polymerase chain reaction, was the same tool that Lipkin, Peiris, and Marra had all used to hook on to fragments of the SARS genome and force it to replicate.) Lipkin had readied himself for the depressing reality that little might remain of most of the samples except a jumble of molecules. But the damage was worse than anticipated. None of the samples was intact, and over three-quarters had been completely destroyed, the wear and tear too great for the fragile macromolecules to handle. This left only about two hundred of the more than one thousand samples that could be amplified, but even these left a lot to be desired: most were just tiny scraps of viral genetic code made up of sections a few hundred nucleotides long. Lipkin nevertheless sifted through the wreckage of the surviving data to find something, anything, that could reveal a link to MERS.

His frustration mounting, Lipkin read and reread the snatches of genomic script that had survived, trying to find pieces that matched sections of the MERS genome that Ron Fouchier had sequenced. For the most part, it was like reading random pages torn out of a magazine: some didn't make sense out of context, while others were like the magazine's masthead, unchanging regardless of the issue. He kept reading anyway, intent on tracking down the pages from the newest issue. And then it happened.

As Lipkin read, a fragment from a single sample, taken from bat feces collected in ruins near the index patient's home in Bisha, stood out. There it was: a hint of the pathogen. The sample came from *Taphozous perforatus*, commonly known as the Egyptian tomb bat, a social creature that roosts in colonies of hundreds, has a taste for moths, and roves across a diffuse habitat stretching from its northern peak in Saudi Arabia, south through Sub-Saharan Africa, and east into the Indian subcontinent. The sample was far from complete—of the thirty thousand or so nucleotide bases that make up the MERS genome, Lipkin was able to salvage only a tiny fraction, a section 182 nucleotides long, representing less than 0.01 percent of the genome. But while the genetic sequence was short, when he compared it against the same section of the MERS virus that had infected the Bisha businessman, he found it was a perfect match.

Lipkin was, if not overjoyed, then at least relieved. As far as he was concerned, it was pretty damn close to a smoking gun. There were reasons to be skeptical, but he was bullish—perhaps too bullish—on the sample. At the time, the scientific hunt to pin MERS on bats had been heating up, with multiple groups publishing studies of more distantly related bat coronaviruses that they claimed were the missing link to MERS. Lipkin hated being outdone. "I looked at their findings," he said dismissively, "and I said, 'Well, that's clearly not the closest match. If that's what's out there, we need to put this out.'" And so he did.

The blowback was immediate. Some scientists pointed out that the specific region Lipkin had sequenced came from a portion of coronavirus genomes that were among the most conserved across the *Coronaviridae* family, meaning that the section didn't mutate very much and could be found in lots of different strains. Lipkin was unbowed. "You'd love to get all four fingers," he said, again likening the samples to crime scene fingerprints. "But you've got half of one finger, and it's a perfect match. That's enough—if you have the rest of the evidence—to say, 'Yes, this was certainly somebody who was here at the scene of the crime.'" Lipkin's point was twofold: First, there was no precedent for a 100 percent match between a bat coronavirus and the index patient's MERS strain. Second, the 182-nucleotide section was just one piece of powerful evidence among many showing that bats were natural coronavirus reservoirs. "This is a bat that clearly has a virus that is identical in this region. So, that's good enough to make that link." Even if the rest of the genome sections were different, Lipkin figured, bats were no doubt carrying, if not MERS itself, then closely related variants that might have spilled over into humans.

There was just one problem. While the Saudis were wrong that there were absolutely zero bats in the kingdom, they were half right in that there wasn't sustained contact between bats and humans. Even a virus perfectly primed for spillover into humans needed sustained contact in close quarters to make the unlikely leap from animals. But Lipkin's own interviews with the Bisha businessman's family and friends confirmed that *Homo sapiens* and *Chiroptera* left each other pretty much alone, and there was no evidence that bats were roosting in places frequented by humans in Saudi Arabia. This meant that Lipkin's tiny, perfectly matched sample wasn't the end of the story; it was the beginning. And with infections, and fatalities, mounting across Saudi Arabia, the pressure was on to find what other strategies MERS had taken to make its byzantine journey from bat to human. Lipkin believed that bats were the key, but others were convinced that another host had helped the virus along its path.

October 2012, Egypt

Saudi Arabia may have been the country where MERS first emerged, but Malik Peiris figured it wouldn't be the only place where antibodies to the virus could be found. Peiris was working on his hypothesis that the new coronavirus hadn't spread to humans in the same way as SARS (through bats and wild animals), but like avian flu: through livestock or other domesticated animals. He knew he might be wrong, but this was how most viruses, especially ones that appeared to be inefficiently transmitted, like MERS, usually gained a foothold in human beings. Avian flu, which crisscrossed the globe effortlessly, had also taught Peiris that epidemics were frustratingly effective at circumventing the borders humanity imposed. If Saudi Arabia wouldn't yield its secrets, Peiris would just have to look elsewhere.

The global epidemiologic network he had built over decades included a Middle East flu surveillance team that tracked viruses across the region. The team had worked in lockstep with Peiris over the years in isolating and controlling avian flu, and he turned to them in his time of need to test his hypothesis that the MERS reservoir would also be found in livestock. First, they fanned out across Egypt, testing humans and animals as they went. Sheep, goats, pigs, cows, water buffalo, and poultry—wherever the team found domesticated beasts, they sampled them. For most animals, it was easy enough to track and sample. For others, it proved to be a much more complicated and bloody undertaking.

Saudi Arabia and its neighbors farmed mostly the same animals as those found in other countries in the world, but one beast, a hardy and truculent one, was the region's crown jewel. Camels were one of the most ubiquitous and highly prized animals in the Middle East, a source of milk, meat, and labor and a revered symbol of resilience and regional pride. They also posed the greatest sampling challenge for the Egyptian team. Peiris figured that the best way to sample camels would be by collaborating with veterinarians, who could broach the idea with camel owners on behalf of the MERS surveillance team. But many refused to take part; camel owners were not particularly keen to have their animals bled. "Some camels are more expensive than humans," Peiris explained. This was an understatement. Beyond mere status symbols, camels could reap their owners millions of dollars on racing circuits and in beauty contests. (In 2018, twelve camels were banned from the annual King Abdulaziz Camel Festival contest, and vying for its $57 million purse, when it was discovered that their owners had injected them with Botox.) It wasn't uncommon for prized camels to sell for upward of $4 million each. While high-end camel ownership was more prevalent in Saudi Arabia than in Egypt, it still proved difficult for the team there to convince owners to part with the blood of their beasts. Eventually, they figured, if they couldn't take blood from live camels, they would have to do the morbid work of collecting it from dead ones. So, the scientists went to abattoirs across Egypt, whose owners were much less fastidious about animal blood. The team hung around, watching and waiting for old, sick, and broken camels to be brought there for slaughter; as the beasts' lives faded, the team scurried to collect an offering from the corpses. "That," Peiris said, "proved a much easier way of doing it."

After being shut out by an entire country and then forced to plead and haggle with fussy camel owners, Peiris and his team saw their tenacity pay off. When they tested the blood and saliva of the humans and animals they had sampled, the results were staggering, not only for what they found but for what they didn't: In the hundreds of samples of humans, water buffalo, goat, sheep, and cows, not a single specimen had antibodies for MERS. But for one creature, the signal was like a glowing red beacon, fiery across the sky; among 110 camels, fully 108 of them, 98 percent, tested positive. Peiris finally glimpsed the epidemic's path clearly. Dromedary camels, a species endemic to the Middle East and North Africa and numbering at least 30 million strong, were a vast unseen reservoir of pathogenic virus hiding in plain sight.

Peiris published the findings of the study on September 5, 2013. Two weeks earlier, Lipkin had been interviewed by *Science* magazine on his mapping of the tiny genetic sequence identical to MERS lurking in an Egyptian tomb bat in Bisha. By then, other evidence had come out suggesting that MERS might be endemic among camels; still, it would take months before Lipkin numbered among the believers. "Those results are interesting, but I'm not persuaded that camels are implicated," he said at the time, his rationale being that MERS patients did not report contact with camels (the Bisha index patient's pet camels notwithstanding). Lipkin still appeared resolutely committed to his bat reservoir hypothesis based on the minuscule fragment he had discovered in a single bat that matched a 182-nucleotide section of the MERS genome. But there were others who weren't as convinced by the famed epidemiologist's conclusions, and they were ready to set the record straight.

MARION KOOPMANS WAS WATCHING THE MERS EPIDEMIC WITH OB-sessive interest. Trained as a veterinary microbiologist and epidemiologist, Koopmans had long been fascinated with emerging viruses.

After finishing her studies in the Netherlands, she had taken a position with the U.S. Centers for Disease Control in the 1990s before heading back to her home country, where she eventually helmed the viroscience department at Erasmus University in Rotterdam. By the time MERS arrived, Koopmans had already faced down multiple avian flu epidemics, which had honed her understanding of how viruses, animals, and humans all found their way into one another's orbits. But that wasn't all: everywhere that zoonotic diseases popped up, Koopmans seemed to be leading the response. She followed outbreaks of norovirus on cruise ships traveling around the world, tracked the ways in which swine flu jumped to humans, deployed mobile laboratories in West Africa during an Ebola outbreak, and worked assiduously to reveal the specific pathways that brought viruses lurking in animals into contact with human beings.

When MERS emerged, Koopmans was in her mid-fifties, a seasoned scientist with an international reputation for being laser-focused and unflappable. No matter the stakes, she could size up what was possible in the moment and, through the power of her convictions, follow paths doggedly wherever they led. This allowed her to soberly weigh various possibilities, regardless of the specifics of the situation, for how a novel virus could end up in a human body. But it also allowed her to weather the inevitable political headwinds that crept into epidemiologic investigations. Though Koopmans wore bright lipstick, and a distinctive shock of spiky platinum-gray lent her a regal, slightly wild quality, her speech was low-key and blunt, sometimes verging on caustic. Still, she had mastered the art of communicating scientific uncertainty in a way that shielded her from the ire that other, more assertive investigators often attracted.

Koopmans had been probing the intersection of animal and human pathogens for her entire career. The veterinarian was a leading proponent of One Health, an approach to controlling new pathogenic threats guided by the basic ethic that animal health, human health, and environmental health are all inextricably connected. It would never be enough for virologists to sequence new zoonotic viruses or for epidemiologists to plan prevention programs.

Viruses, after all, didn't just emerge from a shapeless virome; they were transmitted from animals to humans within specific environments. It was humanity's destruction of the natural world (deforestation, climate change, urbanization, the wildlife trade) that had placed humans and animals in precarious proximity, giving pathogens ample opportunity to move among species, where they probed defenses, adapted, and took hold.

"It is a constant trickle," Koopmans said of the pathogens that, while just beyond the boundary of human contact, were always there, in unknowable numbers, poised to make a jump. "If you change something in the system—if you increase the population of humans, or decrease the population of animals, or create some different way of interacting between species—you have to start thinking, 'What will this do to the pathogens in that system?' Because that's when outbreaks start to happen." Koopmans, who had a tendency to extend her head forward when she spoke, as if searching for the next clue, had seen countless examples of One Health in action.

Koopmans's early work as a veterinary scientist made her particularly attuned to the role camels likely played in shepherding a member of the bat-loving *Coronaviridae* into human society. Though she knew Peiris by reputation, she wasn't aware that he, too, had homed in on camels as a likely reservoir for MERS. Intent on proving the link, Koopmans first did what any self-respecting epidemiologist would do. She tried to establish a baseline against which she could compare her findings. In the case of MERS, this meant finding a herd of camels uninfected with the virus so that she could compare the MERS-negative herd against one in which the virus had successfully proliferated.

The Canary Islands, a tiny archipelago roughly a hundred kilometers off the west coast of Morocco, was an ideal place for her to find MERS-negative camels. Isolated from the mainland and owned by Spain, the islands had long-standing populations of camel that were largely cut off from the rest of their brethren around the world. Koopmans went to the islands and took blood from 105 camels, testing them all for the new virus, fully assuming that she would find no traces of MERS. Wrestling with camels, either MERS-positive or -negative, had its own challenges. "They are mean," Koopmans said. "Camels can be mean." After sampling hundreds of the ornery beasts, she had a piece of advice: "Stay away from the hind legs. Have you ever looked at a picture of a camel and a cow and compared them? The cow has this piece of skin between the belly and the hind legs, which limits its range. The camel doesn't have that—their legs can go fully around. So, they are really dangerous, if you're not on their good side."

After collecting her baseline samples from the Canary Islands camels, Koopmans moved on to the Persian Gulf state of Oman, where she bled another 50 camels, expecting at least some of them to be MERS-positive. In this she was correct, but when she reviewed all her samples, something seemed terribly wrong. Not only was a cluster of the Omani camels positive for active virus, but every single camel she tested in the Gulf state had antibodies for MERS, evidence that they had all been infected at some point in their lives. Shockingly, when she analyzed the samples from her ostensibly MERS-negative Canary Islands camels, the tests lit up: 15 of the 105 isolated beasts also had MERS antibodies. It was an astounding discovery. By all rights, camels in an isolated archipelago shouldn't have been exposed to a virus that had appeared only recently. And yet, there the virus was, circulating among an island herd that had been cut off for decades.

This initial discovery profoundly rattled the investigation of MERS. First, it helped establish once and for all—with further assists from Malik Peiris and Ian Lipkin—that camels were the most likely source of the epidemic. But it also, more disturbingly, implied that the MERS virus spanned whole continents and beyond. The isolation of the Canary Islands camels also made it plain that the virus wasn't a new phenomenon: MERS, or some closely related variant, had been lurking in the humped beasts for years, if not decades, prior to the death of the Bisha businessman.

What struck Koopmans about MERS, though, was its caprice.

Why was it that Arab countries were so much harder hit with MERS cases than countries in neighboring North Africa, or the faraway Canary Islands? After all, MERS was spread by camels, and camels were endemic across all these regions. And yet, hardly any cases of MERS had arisen outside the Arab world. It was almost as if the new coronavirus were picking and choosing its prey, letting some pass unmolested, while forcing others to submit and die. A classic epidemiologist might have focused solely on tracking the epidemic from the time it began to move through human populations. But Koopmans, the veterinarian, believed that the answers lay in that nexus where animal, environmental, and human health collided.

Unlike Malik Peiris's team, Koopmans wasn't interested in slaughterhouses merely as convenient places to collect blood samples. She saw the slaughterhouses themselves—environments where millions of doomed camels met their human executioners—as a critical piece of the epidemiologic puzzle. But she needed to see for herself. In Sudan, the slaughterhouses she visited were actually "slaughter fields," she explained. "Just an open space, and there, and there, and there," she said, gesturing in different directions, "people were slaughtering camels. It was very dispersed, and it was done in the open air, making the risk of getting exposed to the virus quite low." While common across North Africa, these slaughter fields were nothing like the ones in the Arab world. In Qatar, a wealthy Gulf state neighboring Saudi Arabia, the slaughterhouses could not have been more starkly different. Rather than an isolated space to kill camels, the central Qatari slaughterhouse she visited was embedded in a market where camels from across the region were brought together and held as they awaited slaughter or sale. "Camels would be triaged," Koopmans explained: "this one goes for racing, this one goes for milking, and these ones go into the slaughterhouse." As they waited for their fate, the camels would be clustered tightly together by the thousands, often for weeks on end. With the beasts held cheek by jowl, it was inevitable that if MERS was present, it would spread. "It was almost willfully creating conditions for virus circulation."

Koopmans tracked and tested the camels held just outside the slaughterhouse proper as they inched their way inside. Even with protective gear like face coverings, latex gloves, and gowns, it was dangerous work. If she was right, then MERS—which killed between a third and a half of those infected, and for which there was no treatment—would be circulating among the herd. To her shock, 50 percent of the camels waiting to be slaughtered were actively shedding virus. MERS was endemic among the cloistered herd. But it was the building itself that laid the last footstone for the virus to jump from the doomed dromedaries to their human masters. "Inside," Koopmans recalled, "there was a lot of humidity, with constant slaughtering of the animals all day long." Koopmans described a scene of "aerosol and a lot of splashing," as butchers were liberally sprayed with camel blood and saliva. "So," she said, summing up the scene, there are "conditions outside that really increase the level of circulation, meaning that if there's virus, then it can amplify. And then inside, with the aerosol, that's favorable for virus transmission." The Qatari slaughterhouse was a coronavirus factory, efficiently churning out live virus from the dead and dying camels. Koopmans had found what epidemiologists had longed for: the uncapped source of the epidemic.

November 2013, Saudi Arabia

Koopmans and Peiris were able to pinpoint, beyond a shadow of a doubt, the path that MERS had taken to infect human beings. Camels, it turned out, were not only mean but also the reservoir for the deadly coronavirus. Among the convinced was Ian Lipkin. As the only researcher with ready access to Saudi Arabia, Lipkin realized it fell on him to return to the kingdom with his team, this time with a new mission: to stop the transmission of MERS from camels. Unlike with bats, nobody could deny that camels were anything but ubiquitous across Saudi Arabia, so Lipkin could quickly get to work testing the animals. Within two months, his team had taken blood, saliva, and rectal swabs from about two hundred camels across the country. Here, finally, were his five fingerprints, or so it seemed: fully three-quarters of camels over the age of two tested positive for MERS antibodies, though few actually contained live virus. Among the younger camels, though, Lipkin's team was able to extract entire virions and large fragments of MERS genomes. Remarkably, when Lipkin tested old blood samples taken prior to the outbreak of MERS in 2012, he found that camel blood from as far back as 1992 tested positive for MERS variants. The blood told the story of an endemic virus, of widespread circulating infection among young camels, and of a long-standing viral reservoir that had been churning, drifting, and shifting for at least a decade, evidently without a pathway into human beings before the businessman from Bisha succumbed to one of its progeny.

When Lipkin's study was published, he thought it might spur renewed urgency from the Saudi government to contain the apparent animal source of MERS. In this, he was sorely mistaken. If mentioning bats was taboo, criticism of the camel—the official animal of Saudi Arabia, a source of deep national pride, and a major money-maker for the kingdom—was an act of scientific aggression.

In a sharply worded response to his paper in *mBio,* the same peer-reviewed journal in which he published his findings, two scientists from King Saud University in Riyadh tried to pick apart, piece by piece, Lipkin's conclusions. It was no coincidence that the scientists hailed from the Department of Animal Production at the College of Food and Agriculture Sciences, a global hub of camel research that hosted projects with names ranging from "Genetic diversity of camels, sheep and goats in Saudi Arabia" to "Camel nutrition and ways of improvement." Lipkin had pissed off the camel scientists, and the mounting evidence of camel-to-human MERS transmission was messing with a steady source of profit. It wasn't just the upmarket camels destined for racing and beauty contests, literally in a breed of their own. The kingdom was also dependent on a massive camel meat market, estimated to be worth as much as $750 million annually, the largest in the Middle East.

Trying to debunk Lipkin's work was an uphill battle. It was impossible to take issue with the science itself—Lipkin's team was among the best in the world at sampling, viral purification, and genomic sequencing—so the camel scientists tried a different tack. Sure, they conceded, most of the camels had tested positive for MERS. But that in itself wasn't proof that the animals were a natural reservoir for the virus, which Lipkin was claiming. Assuming as much, they chided, amounted to breaking a cardinal rule of epidemiology: speaking beyond the data. While it may have been plausible that camels were responsible for MERS's zoonotic jump to humans, Lipkin's results had by no means proven it beyond a shadow of a doubt. In fact, the scientists argued, what Lipkin saw as a camel-to-human pathway might actually be the reverse. "Human-to-camel MERS-CoV transmission is also of importance," they wrote, going on to suggest a scenario whereby MERS jumped from cats or some other animal into humans, who then passed it on to its real victims, camels. "We can clearly see that it is plausible to assume that humans (or any other source) are the ones who infected dromedary camels in the first place." It was a fabulist interpretation of the data, to be sure, though beyond its absurdity lay a real fear: The camel industry had been decimated as the beasts became the prime suspects behind the MERS epidemic. If Lipkin, Peiris, and Koopmans were to be believed, MERS was endemic among camels, meaning that every sneeze or cough from a camel—and there were many— put humans at risk of becoming infected and, more likely than not, dying of MERS. Already, governments in the Levant, along with the World Health Organization, had released recommendations warning people to exercise caution around camels and not to consume raw camel milk or meat. It was a potential death blow to the camel industry, and it was scientists like Lipkin who had dropped the hammer. Even as a pathogenic human coronavirus emerged out of the hidden virome for the very first time, Lipkin's was a familiar story. Money and politics were always at play when nature retaliated against our species.

Notoriously circumspect though the Saudi government was, there were signs that it was split on how best to handle MERS. Senior figures like Dr. Ziad Memish, the deputy health minister, were motivated to generate scientific findings on the causes and consequences of MERS. The virus, though, had evidently wounded Saudi pride while threatening to upend one of the kingdom's most valuable industries. It also posed risks for the upcoming hajj, the annual Islamic pilgrimage to Mecca, which was on Saudi Arabian soil and which attracted more than two million penitents. These competing interests—an instinctive desire for autocratic control tempered with a recognition of the severity of the epidemic—were on full display as the Saudi government sought a path through the chaos. Though it took the better part of a year, by the fall of 2013 the factions within the kingdom pushing for the virus to be taken seriously appeared to have won out. The government eventually extended warnings about consuming camel meat and milk and worked to enact protocols to stop the coronavirus's spread among humans.

Lipkin was keen on helping the Saudi government with its change of heart. Unlike the Chinese government during SARS, though, Saudi officials evidently now saw him with less than affection. Lipkin realized that if he was going to get back into the country, he'd have to make his own invitation. His solution was to leverage his relationship with the financial giant McKinsey, which arranged to have him invited back to local Saudi hospitals to improve their MERS-prevention protocols. McKinsey, though an ostensibly odd entrant into the fight against the deadly coronavirus, had a long-standing relationship with Saudi Aramco, the massive oil conglomerate owned by the Saudi government, which was worth over two trillion dollars. To hear Lipkin tell it, McKinsey's financial relationship with Aramco allowed it to dictate terms across a range of issues in the kingdom, including, apparently, those as far afield as responding to novel coronaviruses. This was no surprise to those in the know. The global consulting firm's influence on Saudi affairs was such that the kingdom's Ministry of Planning had long been nicknamed "the Ministry of McKinsey." If McKinsey wanted Lipkin in Saudi Arabia, it was as good as done.

What Lipkin found when he returned a year later was that, like

SARS, MERS was mostly spreading among frontline health workers. It was the intensivists (doctors who cared for patients in the ICU and whose skills included up-close intubation) who were contracting the virus at an alarming rate. What puzzled the Saudis was that the intensivists were trained in infection control and had taken all precautions, including donning high-quality PPE like masks, gowns, and gloves. And yet, the virus was still doggedly finding its way into their lungs.

It didn't take long for Lipkin to pinpoint the cause. The intensivists' devotion to epidemic protocol, and to the wearing of standard-issue N95 masks (so-called because they protect their wearers from a minimum of 95 percent of the dust and droplets circulating in the air), was unwavering. Equally unwavering, though, was their devotion to Muhammad, the one true Prophet of Allah, who had commanded his acolytes to "cut the moustaches short and leave the beard." In Saudi Arabia, a deeply religious country, the beard was a critical part of Islamic identity. Lipkin realized that the intensivists' beards were making it impossible for them to form a seal between the skin of the face and the N95 mask itself, which fatally weakened the mask's protection. Without the seal, respiratory droplets from MERS-infected patients could still find their way into the intensivists' mouths by clinging to their facial hair and slipping in through the slim chasm between their skin and the mask's polypropylene fibers.

Asking men in Saudi Arabia to shave their beards was out of the question. Lipkin instead proposed that the bearded clinicians switch from N95 masks to more cumbersome but fully sealed powered air-purifying respirators (better known as PAPRs), cowl-like apparatuses that enclosed their entire heads, creating a positive pressure environment around their faces that made it impossible for pathogenic droplets to get anywhere near their airways. It worked. After Lipkin's intervention, cases of MERS among clinicians dropped substantially.

What was more difficult, though, was undoing the deep inequity that existed in Saudi Arabia, which provided ongoing human fodder for the virus. The divide between the kingdom's residents and its estimated nine million guest workers, many hailing from East Africa and Southeast Asia, had manifested itself in many ways. Some observers, like Human Rights Watch, described the condition of domestic servants in Saudi Arabia as "near slavery." Meanwhile, public beheadings of migrant workers for crimes such as "sorcery" weren't out of the question. There were more subtle abuses, too. Guest workers were regularly denied access to hospital care, and when they were admitted, they were unceremoniously forced into waiting rooms overflowing with immunosuppressed people who stood tightly packed together, nose to nose, an underclass forced to endure both daily humiliations and exposure to pathogens.

By 2015, THREE YEARS AFTER THE DEATH OF THE BISHA BUSINESSMAN, in the wake of numerous investigations into the origins of MERS in Saudi Arabia and across the region by the world's greatest living coronavirus epidemiologists, the world was still largely in the dark about key facets of the origins of MERS. Beyond the fact that camels were involved, there was no scientific consensus on how the virus was transmitted from animals to people, and between people, across the Arab world and beyond.

What MERS made crystal clear, for those who had been following coronaviruses since before the advent of SARS, was that members of the *Coronaviridae* were more than a one-off threat. While SARS carried a 10 percent fatality rate, MERS killed roughly 35 percent of those it infected. The two viruses, from different branches of the viral family, had emerged a decade apart. They were twin events far beyond the realm of chance. The family, long seen as innocuous, had, in the blink of an eye in viral time, proven that it was a keen and adaptive hunter.

And still our species did not, by and large, heed the glaring, ominous warnings. There were some, though, who quickened their preparations for the coming catastrophe.

Chapter 7

IF YOU'RE A MOUSE, THE ANSWER IS PROBABLY YES

N O MATTER WHAT THE NUMBERS SAID, RALPH BARIC HAD ALWAYS been adamant that SARS was a rare event. Yes, there were potentially tens of millions of unique coronavirus variants circulating among bats, but it was still exceedingly unlikely for any new virus to emerge out of the zoonotic soup, find its way into a human host, and then, miraculously, begin replicating efficiently enough to infect thousands of people. Moving between species, even for a well-adapted virus, was an impossible journey. Perhaps it wouldn't have been so surprising in an inconstant virus known for its extreme flux, like influenza. But the one thing everyone had always agreed on was that coronaviruses were predictable, slow-evolving microbes. When Baric first learned of them, they were a puzzle for him to solve on his own terms, a portal through which to explore the liminal space between life and molecule. Yes, he admitted, SARS had shown that even the most seemingly benign virus could slip from the hidden virome and into human society, bringing chaos. But that was why he had declared himself shocked but not surprised by its emergence. The numbers bore out the possibility, though a remote one, that a coronavirus could alter itself sufficiently to cause an epidemic. Even so, in the wake of SARS, most observers, Baric included, did not consider coronaviruses as the top pathogenic threat humanity might next face.

When Ali Zaki's call went out across ProMED announcing that a novel coronavirus had killed a man in Jeddah, Baric's views shifted abruptly. It wasn't that humanity should have known something it did not. Instead, Baric, as he so often did, reconsidered the intrinsic nature of the virus he had spent thirty years investigating. He could name from memory each section of its genome and run through the idiosyncrasies of each strain's structure down to its individual bases. The particular nucleotides in the two open reading frame sections of their genomes or the specific glycoproteins that made up the receptor-binding domain that unfolded like flower petals at the end of the spike protein—Baric had probed them for ineffable truths about the *Coronaviridae*'s relationship with its hosts and how its members propagated their pallid version of life. But MERS revealed to Baric that he had been thinking about the family all wrong.

Though he had no clear answers, he had something even more valuable for a scientist: a question. "A new virus emerging from animal reservoirs and causing eight thousand cases—that's a rare event," he noted, referring to SARS. "And then to have it happen again in 2012 . . . That made me more interested in asking whether the animal reservoirs were filled with generalists." Instead of a viral family steadily honing its adaptation to a specific species, the quick succession of SARS and MERS—which occurred almost simultaneously in the universe of viral time—suggested to Baric that coronaviruses were evolving in the other direction, to infect an increasing number of hosts. The question raised disturbing possibilities. "If they are generalists, how big of a threat is this to human populations?" he asked. "How do we best get prepared to combat that? Because if two emerge in less than ten years, something different has happened in the ecologic landscape. And we need to seriously begin to think about how big the threat level is and how to prepare for it."

Baric was arguably better prepared than anyone in the world to know what questions to ask. Still, he wasn't omniscient. Though he represented the greatest store of knowledge on the peculiar intricacies of coronaviruses, what he knew was dwarfed by the colossal mysteries that the *Coronaviridae* withheld from view. Case in point:

though he and others knew what functions most of the sections of the coronavirus genome carried out, they didn't know how novel mutations at specific places, and which specific bases, would affect the integrity of the viruses. They also had only a fuzzy idea about what some of the seemingly less important parts of their viral structure actually accomplished.

This molecular myopia notwithstanding, Baric did have a sense of the ways in which the viral family might yet evolve to overpower humanity. These included the most distant, rarest, and scariest outliers of what a future zoonotic leap could unleash. The worst, in short, was a pandemic-ready pathogenic coronavirus. With that as a real possibility, inaction was not an option, though it wasn't entirely clear what that meant when the virus you were trying to protect humanity from didn't yet exist. Baric, though, wasn't satisfied with simply sitting on his hands until better information came along. The stakes were just too high. "At a certain point, you have to decide what kind of questions you choose to pursue," he said. "You don't have unlimited people; you don't have unlimited resources. So, you pick and choose."

It wasn't just people and resources that hemmed Baric in. It was also the political environment around virological research that he confronted in the months after the emergence of MERS. In the past decade, the world had not only seen two new pathogenic coronaviruses, but had also faced a resurgence of H5N1 avian flu, first detected in 1997, that had spread across the globe. By 2013, avian flu was still causing sporadic human outbreaks and millions of bird deaths. In response, flu virologists had undertaken transmission studies, wherein they probed how close a flu virus might be to jumping from animal to human and which mutations were needed to get it there. These sometimes involved selecting for highly virulent flu strains in the lab, the experiments forcing the virus to adapt or die in the face of a new challenge. If the viruses being manipulated were already dangerous to humans, this posed an outsize risk for the scientists doing the work—and for the world beyond their laboratory walls. Still, it was a calculated risk. The principle was that transmission studies gave humanity a glimpse at the weaponry that the virus might bring to bear and a sense of how quickly it might evolve to attain it. That could provide an early advantage in an outbreak by allowing researchers to proactively work on a vaccine, therapeutics, and effective public health protocols to restrict infection.

Nobody, though, had bothered doing transmission studies of coronaviruses, which were (SARS and MERS notwithstanding) widely considered stable and benign. Baric considered the possibility. Unlike the flu, coronaviruses mostly used mutations in a specific region of their genome (the highly mutable ACE2 receptor-binding domain, the flowering buds at the tip of the spike protein) to unlock the cells of many different species. But mutations that made a coronavirus more efficient at infecting one species generally made it less efficient at infecting others. This meant, in principle at least, that coronavirus transmission studies in animals should be relatively safe to do in a lab because if a highly transmissible virus somehow got loose, it wouldn't be well adapted enough to infect humans. This was just a hypothesis, though. "The truth is," said Baric, "I don't know."

Beyond the logistics of running the experiments, Baric was worried about the political arena outside the lab that confronted him in 2013. "The National Institutes of Health, the World Health Organization, the FDA, the CDC—they all said the most important question in flu pandemic preparedness was to understand which mutations drive cross-species transmission and increase transmissibility." This had led virologists to take existing avian flu viruses and modify them to be as virulent as possible. "Two labs did that," Baric said, "because research funders wanted to figure out how easy it was for avian flu to learn how to transmit. They figured it out. But that's an edgy experiment, and they took a tremendous amount of grief for it—doing exactly what the governments of the world asked them to do." That grief, which Baric recalled vividly, came in the form of howls of protest from concerned scientists across the globe, followed by a cascading set of scandals around scientific ethics, free speech, and pandemic preparedness.

One of the scientists at the center of the firestorm was Ron

Fouchier, the same Dutch microbiologist who first sequenced MERS after Ali Zaki sent a viral sample from the Bisha index patient and who worked with Marion Koopmans on tracking viral outbreaks across the globe. Though he dabbled in coronaviruses, Fouchier was primarily a flu specialist. In 2013, H5N1 was still spreading very slowly among humans, having infected only six hundred people since 1997. But 60 percent of those infected had died, and fears that a mutated version of bird flu might transmit more efficiently while maintaining its deadliness spurred Fouchier to action. In his lab at Erasmus University in Rotterdam, he tried to coax into existence a version of H5N1 that could rapidly infect ferrets, a mammal with a similar immune profile to humans. The goal was to get the virus, which could not spread between the animals through the air, to learn to fly. Fouchier's experiment was wildly successful. After he created just ten artificial transmission events (by rubbing mucus from one infected ferret into the nostrils and mouths of the others), five minor mutations appeared in the virus's genome. This antigenic drift allowed Fouchier's lab-based bird flu to infect ferrets through the air—while still remaining as lethal as ever. The implication was clear: H5N1 was perilously close to being the world's next major pathogenic threat, and we needed to immediately begin to prepare for it. Fouchier had done what he had been asked by his funders to do: he had sounded the alarm.

Others saw things differently. Since 2001, when anthrax was mailed to prominent figures in the United States in the chaotic weeks after the 9/11 attacks, ultimately killing five people, the U.S. government came to see pathogenic viruses primarily as a biosecurity threat rather than a public health issue. In the paranoia and ballooning security apparatus that marked the years after 9/11, the U.S. government established a National Science Advisory Board for Biosecurity, with a mandate to ensure that the risks posed by emerging pathogens, both those spawned by nature and the grotesques artificially produced in labs, never threatened humanity again.

When Fouchier went to publish the results of his flu transmission study in *Science,* the Biosecurity Board stepped in. The data, they determined, were too dangerous to be shared publicly; Fouchier had basically written a recipe for producing a bioweapon. Instead, in an unprecedented move, the board demanded that Fouchier and the journal redact key parts of the data, including the specific location of the mutations that transformed H5N1 into an airborne killer. Fouchier ultimately agreed, though he clearly thought it was overkill. He pointed out that by the time the data were published, it would already be too little, too late: more than a thousand people, including his lab colleagues, collaborators close and far, the journal staff, editors, and anonymous peer reviewers had already learned the location of the mutations. "As soon as you share information with more than ten people," he said in an interview at the time, "it's no longer confidential." Fouchier's solution was to advocate for the continuing need for "dual-use studies" (so called because their results could ostensibly be used for good or evil), while publishing results in what he called "a responsible way." Others disagreed. If, like pathogenic viruses, dangerous data found a way to spread, then maybe the experiments shouldn't be conducted in the first place.

Surveying the wreckage wrought from good intentions, Baric made up his mind. "I had no desire to get into that arena," he said. For a scientist whose career began with the decision to study coronaviruses instead of HIV, it made all kinds of sense. Baric didn't need the grief that came from transmission studies. He just wanted to continue working in his lab doing interesting experiments. That's not to say that he didn't see the good. "Was that experiment of value? I think it was, because it identified mutations that could be associated with a high-path avian flu pandemic. Was there risk, though? Yes, you created things in the lab that had not yet occurred in nature. And you've produced a blueprint for how you did it."

Baric took issue with experiments that made dangerous viruses even deadlier, but he wasn't averse to using the same techniques as Fouchier. Baric's rule was that he would create synthetic pathogens to advance knowledge, but he would steer clear of creating more dangerous ones than those that emanated from nature. He believed that the ethical blowback against dual-use experiments shouldn't extend to every scientist who was manipulating viruses in the lab; evidently, he saw himself as one of those who should be exempted. Baric had started artificially recombining different coronavirus genomes in the mid-2000s, when he reverse-engineered the SARS variants that diverged at each of the waves of the epidemic. In that case, he wasn't using synthetic recombination to produce wholly new viral strains but to reconstruct early, less transmissible versions of SARS to understand how the pathogen eventually managed to spread through humans. Artificially designing badly adapted viruses was a much less ethically fraught experiment for Baric than designing ones that could very well cause pandemics.

Still, he conceded that it was a question of degrees. "When you take the spike gene of one strain and you put it into the backbone of another strain"—something he had been doing for decades, mostly by grafting the spike proteins of human coronaviruses onto mouse coronavirus—"theoretically, that does not exist in nature. So that would be a synthetic virus." This was known as a chimera, and while it might sound scary to the uninitiated observer, in the spectrum of artificial viral life forms, Baric saw in his brood low risk and high reward. He was taking human coronaviruses and making them less deadly by adapting their most important features onto mouse-adapted strains. In the unlikely event of an escape, mice were the only creatures that had anything to fear.

There was another dimension to Baric's decision to create chimeras in his lab. It was, in short, money. Since the 1990s and with the advent of genomic sequencing, there had been a brisk market for synthetic DNA. Every genetic sequence published in the publicly accessible genome bank run by the National Institutes of Health could, in principle, be used as a blueprint for constructing real-life viruses. In the 2000s, the price of a synthetic genome thirty thousand bases long (roughly the size of coronavirus RNA) was fifty thousand dollars, a massive capital investment for virologists working in a still-obscure discipline hardly awash in funding. By 2012, while the price had gone down to about ten thousand dollars for a full coronavirus genome sequence, it was still a high enough barrier to entry that it forced scientists like Baric to get creative. "As an investigator you're in this quandary," he explained. "You've got eight bat viruses that you think might be interesting to work on, but you only have ten thousand dollars. So, you have to pick just one." Faced with this dilemma, Baric would order the full sequence of the coronavirus strain that looked the most interesting, which was often the one that had a spike protein efficiently adapted to infect a cell. Then he would synthesize that spike protein and recombine it with mouse coronaviruses in his lab so that he could test how each performed under pressure. With more money, Baric suggested, all this cut-and-paste genetic work wouldn't be necessary. He could just pay to synthesize all the sequences he thought worthwhile to experiment upon. In any event, he reasoned that the recombining of coronavirus genes he did in the lab could never outpace the torrid speed of natural evolution. "Recombination occurs at a very, very high frequency in some coronaviruses," Baric said. "And so anything that is a chimera, that is made in the laboratory, has probably been made in nature as well."

While born of financial necessity, synthetic recombination became the backbone of Baric's work through the late 2000s and the 2010s. It was a vicious circle. The more damage that human coronaviruses caused, the greater the impetus for field epidemiologists like Shi Zhengli to harvest as many new coronavirus strains as possible. As more strains became identified, a more detailed picture of the extent of the coronavirus family came into view, which gave all the more reason to bring the new viruses into the lab and test how dangerous they might be.

BARIC WASN'T A SHOE LEATHER EPIDEMIOLOGIST, OUT IN THE FIELD scratching bat guano off stalagmites in fungus-encrusted underground caves. That work, undertaken by scientists like Shi, maintained a steady flow of data that he used to reveal the guile coronaviruses employed to leap between species. But with his prowess in probing the structural biology of viruses, and Shi's aptitude in

collecting them, it was only a matter of time before these two pow-
erhouses crossed paths. Baric had been following Shi's investigations for years, though always from a distance—evidently, he wasn't
the spelunking type—and he had become alarmed by what her field
research had uncovered in the defunct Tongguan mine and the vast
Shitou cave complex in Yunnan Province. Shi, Baric well knew, had
captured hundreds of coronaviruses in bats, enough to confirm that
the flying mammals were a natural host reservoir for multiple members of the *Coronaviridae*. But in 2013, Shi's team published truly
alarming findings. The Yunnan bats she had sampled, in a region
that hadn't recorded a single SARS case and that lay almost a thousand miles away from Hong Kong, were carrying a potentially
terrifying new coronavirus strain. Dubbed RsSHC014 (after Shi's
taxonomy using bat species type, location, and the sample number),
the strain was 95 percent identical to SARS, much more closely related than any other bat coronavirus that Shi or anyone else had
previously discovered. RsSHC014's similarity to SARS could mean
one of two things: either it was a pandemic-ready pathogen that
needed only a chance collision with a human host to start the cycle,
or the 5 percent difference between it and SARS was sufficient to
make it effectively inert. Either way, Baric figured that RsSHC014
was the best chance he had to test whether coronaviruses were the
hardy generalists he feared they might be.

The most important data point Baric had at his disposal was
SARS, which had successfully spilled over into humans and then
caused epidemics in every region of the world. (While MERS had
also spread to twenty-seven countries, it was a far less efficient transmitter.) In the absence of any other revelatory information, this
made SARS-like viruses, and RsSHC014 specifically, the most
likely suspects in a future zoonotic leap. Despite all that he had
learned over the decades, he, like everyone else, was stuck trying to
anticipate the next pathogenic threat with no idea what to look for.
It was a bit like trying to find a cliff face on a mountainside by walking backward in its general direction. All you could do was study the

terrain you had already traversed and hope that the patterns in the gravel would be enough of a warning before you plunged to your death.

Within months of Shi's discovery of RsSHC014, she and Baric had struck up a collaboration. Over written correspondence, Shi agreed to send the spike sequences of RsSHC014 and other sections of its genome to Baric's lab at the University of North Carolina in Chapel Hill. Baric's team, mice at the ready, would then take over.

The series of experiments Baric undertook on RsSHC014 were like an extended high-wire act. Minutely probing the vast, undiscovered terrain of the virus, he simultaneously created a simulacrum of the way humans would respond to it. He wanted to know not only how RsSHC014 would exploit weakness in its hosts, but how the arsenal of human weaponry designed to protect us from SARS would measure up against a closely related pathogen. The first order of business was to take the RsSHC014 spike and synthetically recombine it to a mouse coronavirus strain that, before it was altered, didn't make the animals sick. That way, he could understand how important the RsSHC014 spike protein was in transforming a benign virus into a pathogenic threat. To quantify how worried humanity should be about the new SARS-like bat coronavirus, Baric then introduced it directly into human cells, to see whether the virus was already able to infect humans without needing to mutate further. It was his attempt at re-creating that first phase of a coronavirus epidemic, when the novel pathogen made the leap from animal to man.

Still, there was more to do. Baric's BSL-3 lab was stocked with mice, tissue cultures, viral samples, and reams of PPE and redundant fail-safes to protect those who worked inside (and those who lived outside). Baric had also had monoclonal antibodies taken from patients who had been infected with the original SARS virus and had survived. He figured that, in the absence of a vaccine, these antibodies were humanity's next best hope for a cure if the virus reemerged. But no one knew whether antibodies to previous coronaviruses would protect us from future ones. RsSHC014, which was 95 percent identical to SARS, gave Baric the perfect opportunity to mimic exactly that situation.

There was one last step, and for Baric it was arguably the most critical. Over the previous decade, many scientific teams, including Bob Brunham's, had been developing SARS vaccines that hadn't ever advanced to market because humanity didn't need a SARS vaccine after the virus was driven back into the wild. Nevertheless, preclinical SARS vaccine trials that Brunham (in the SAVI era), Baric, and others had run in animals had shown promise. So, Baric asked, if a SARS-like coronavirus caused a pandemic, would those existing SARS vaccine candidates be worth anything?

While all four of these steps—mouse infection, human cell infection, antibody testing, and vaccine testing—were useful on their own, Baric's scientific elegance shone through once the lens shifted out and they were considered as a whole. It was then that dryly reported results in a scientific study became a gripping narrative of the next coronavirus pandemic played out in miniature. Under Baric's omniscient manipulation, RsSCH014 illuminated a possible pandemic future: a coronavirus emerges in the hidden bat virome, then recombines and enters another ubiquitous mammal, before spilling over into a human being. Once ensconced, it burrows its way into its new host's cells, seeking pathways to infect and replicate. The threatened host mounts an immune response (antibodies), hoping that its biological system's weaponry is nimble enough to blunt the infection. And finally, a vaccine is developed (most likely on the back of preexisting technology) that can, one hopes, end the pandemic once and for all. Baric's experiment was a multidimensional exploration, balancing the microscopic (deep probes of the molecular structure of coronaviruses) with a rarefied sense of a pathogen's grand advance across the human race. With the knowledge he gained, he could shift his gaze over his shoulder and glimpse how close the cliff face really was and how far down the drop might be.

Baric had many fears. He worried that SARS and MERS were only the first in an accelerating series of novel pathogenic coronaviruses. He worried that humanity was totally unprepared for what would come next, when a truly fit coronavirus emerged. But in the very short term, he worried about attracting undue attention. Starting with choosing coronaviruses over HIV, and then avoiding transmission studies like the one Ron Fouchier ran on H5N1 avian flu, Baric was convinced that any hint of controversy would slow down his work, to the detriment of all. In 2014, poised to expose just how close to the pandemic precipice RsSHC014 had brought humanity, and with the stakes as high as they had ever been, he was surer than ever that he had made the right choice. Outside the discipline of coronavirology, where he was a titan, nobody really knew who he was. It was exactly what he wanted. Like the *Coronaviridae*, the workings of Baric's lab remained largely oblique, the better to allow him to advance humanity's knowledge in peace.

But perhaps Baric was too sure. Despite his dreams of being left alone in the universe of coronaviruses into which he had disappeared for over thirty years, the human world finally caught up to him.

October 24, 2014, Raleigh, North Carolina

It was one of the best days of Baric's life. Surrounded by his family and friends, the patriarch of the Baric clan surveyed the gathered group with unfettered joy. With his wife and children by his side, he watched as his daughter, radiant as ever, dressed in white and exuding boundless happiness, exchanged vows with her new husband. For a moment, viruses didn't matter.

Two days later, on a sunny Monday morning, Baric entered his office still wrapped in the warmth and goodwill of the weekend. As he turned on his computer and scrolled through his emails, the feelings of paternal joy abruptly drained away. There was a letter in his in-box forwarded from the senior administrator at the University of North Carolina at Chapel Hill, sent by the U.S. Department of Health and Human Services, and titled "Re: 5U19 AI107810-02." Baric's heart sank. The obscure string of numbers was the ID for the coronavirus research funding, worth millions, that he had received from the U.S. government. "Dear Ms. Settle," the letter began (addressing the senior administrator), "the National Institute of Allergy and Infectious Diseases has determined that the above-referenced grant may include Gain of Function (GoF) research that is subject to the recently-announced U.S. Government funding pause, issued on October 17, 2014." The letter continued, specifically naming "Ralph S. Baric, Ph.D." and demanding that the work he had been doing to test "novel functions in virus replication in vitro" and "novel functions in virus pathogenesis in vivo" be immediately paused until further notice. Despite his assiduous efforts to avoid attention, Baric had been caught in a net.

Gain-of-function experiments test whether a pathogen (or potential pathogen) can evolve new traits under pressure. Baric's work from the 1990s, when he was able to transform MHV, the mouse coronavirus, into a hamster-adapted virus, was a classic example of gain-of-function research; so were the dozens of experiments he had run over the decades in which he had synthetically recombined parts of SARS, MERS, and other coronaviruses with MHV, which allowed him to see how those pathogens mutated to adapt to new conditions. Baric's experiments appeared, to the uninitiated observer, audacious; they were also, in the spectrum of gain-of-function studies, fairly innocuous. When asked, during an NPR interview conducted after the shutdown announcement, whether he was making pathogens more dangerous, Baric responded tartly: "If you're a mouse, the answer is probably yes, or at least I was trying to." It didn't take much reading between the lines to understand that he saw his experiments as falling far below the threshold of what constituted a serious threat.

The pause directive had originated with the same Biosecurity Board that had been anxious about Ron Fouchier's gain-of-function experiments, which had transformed avian flu into an airborne killer three years earlier. By 2014, things had come to a head when high-profile protocol breaches at U.S. CDC labs thrust the potential dangers of these studies into the spotlight. Whistleblower leaks revealed multiple allegations of laboratory carelessness with materials used in risky experiments. In one incident, dozens of CDC workers were potentially exposed to anthrax after a routine experiment went wrong. In another, vials of smallpox were left lying around unguarded in an open storeroom. In yet another, vials of seasonal flu were shipped across the country after being contaminated (without the researchers' knowledge) with a deadly form of avian flu. These were just a few among many incidents. Over the past decade, it was revealed, pathogens at labs across the United States had been mishandled, lost, or stolen on average twice every week. It was a terrifying pattern of negligence with the potential to unleash a wave of death.

This had prompted two hundred concerned scientists, Ian Lipkin among them, to release in July 2014 a "Consensus Statement on the Creation of Potential Pandemic Pathogens," sternly warning that a course correction was needed before the next pandemic emerged from a lab leak. Two weeks later, another hastily assembled group of scientists calling themselves Scientists for Science, which included Ron Fouchier, Susan Weiss, Baric's collaborator Mark Denison, and others, responded with their own statement. (Baric, true to form, stayed out of the fight.) There were enough regulations already, the Scientists for Science argued. And anyway, what was the point of having biosafety level 3 and biosafety level 4 labs if you weren't going to use them to study dangerous pathogens? Though the defenders of gain-of-function studies were impassioned, they were ultimately unconvincing. When Baric received the NIH directive to halt his research, it was clear that the Scientists for Science had lost the battle.

Baric had his own doubts about experiments like Fouchier's, which he described as "edgy." Still, he was surprised that he had been dragged into the fray. As far as he was concerned, his research was miles away from what Fouchier was doing. The only pathogens Baric manipulated were synthetically grafted onto a mouse coronavirus, which he believed put the risk that a pandemic-ready pathogen could escape from his lab at close to zero. Still, it wasn't always easy for those on the outside—even other virologists—to keep up with Baric's seemingly endless thread of ideas and experiments or to gauge their risks. For a public unfamiliar with the intricacies of his work, the talk of synthetic recombinants, hybrid SARS-like viruses, and chimeras conjured images of scientists running amok like in the H. G. Wells novel *The Island of Dr. Moreau*. It frustrated Baric to be lumped into the same category, given how assiduously he had toed the line, not to mention how high the stakes were. "There's risk in BSL-3 work and recombinant DNA research," he admitted. "There's risk to the individuals, and there is some risk to society." The question for him was how prepared humanity wanted to be for future threats versus how reactive in the moment. "Should we have people living in the present prepare for catastrophic events in the future? The answer to that should be yes, be it global warming, a pandemic virus, or asteroids that can smash into the earth." The topic raised his ire. "These are smart things to do," he said. "We're in it for the species."

The cruel irony was that the 2014 gain-of-function shutdown, which had the intended and actual effect of seriously slowing down research on emerging pathogens, came right when the world could least afford it. MERS had, two years earlier, shown that SARS was no rare event. This was followed by Shi Zhengli's discovery of a massive cache of bat coronaviruses, which Baric calculated might include as many as ten million variants. There were also Baric's experiments suggesting that those in the *Coronaviridae* family were generalists, able to infect multiple species, along with his research revealing the elegant programming of the spike protein, which made coronaviruses poised to adapt to multiple species, including our own. And then there was the accelerating level of interaction between animal reservoirs and human beings that Marion Koopmans had documented across Qatari camel abattoirs. All pointed to the same unavoidable conclusion: the era of pathogenic coronaviruses was only beginning.

But when the National Institutes of Health tells you to stop running experiments, you comply. Baric, as usual, was immersed in multiple projects at once—including the tests he had planned to run on RsSHC014—and was forced to shut many of them down and sit on his hands. He was barred from starting new experiments with MERS and SARS. While he understood the logic, it was nonetheless deeply frustrating. He had worked diligently on a model of pathogenic experimentation that minimized the risks to humans. All that was now for naught.

The pause reverberated deeply across the landscape of virology. Until 2017, when the moratorium was finally lifted, scientists were prohibited from any new experiments on flu, SARS, or MERS that might imbue the viruses with new traits. This had a predictable effect. The study of pathogenic viruses stultified, and so did the field at large. In 2015, 909 coronavirus studies were published, the highest number ever; in 2016, there were 884, the vast majority using data preceding the gain-of-function pause; in 2017, there were 777, as the last remnants of pre-pause data finally made their way into peer-reviewed journals. And then, in 2018, there were only 112. The field, already small, was shrinking, its output declining. It made sense; why would an up-and-coming virologist study coronaviruses when they wouldn't be able to wrestle with the family's most dangerous, and important, members? The field, largely left in the hands of scientific elders, appeared to be slowly dying out, despite the undeniable evidence that the collision of coronaviruses with our species was accelerating.

As soon as Baric got word of the pause, he began to fight back. His laboratory was immersed in the most consequential phase of coronavirus research he had embarked on in three decades; stopping it cold was just not acceptable. He had experiments testing how similar the immune response was to MERS, SARS, OC43, and NL63, which was critical to understanding what the next coronavirus strain would look like and how the human body might react to it. He had vaccines and antivirals against MERS to test, which he could do only with the virus itself (or a chimeric version of it designed to attack mice). And then there was Shi Zhengli's trove of bat coronaviruses, chief among them the SARS-like RsSHC014, which could have occupied a virologist for decades. And in the background, as always, was the specter of the next spillover event. Baric had to get back to work. For weeks, he pleaded his case, to no avail. The NIH was firmly against starting any new research that could be classified as gain of function. But there was a caveat, he was told: the pause wouldn't apply to experiments he had already started. Baric breathed a sigh of relief. Thankfully, he was a scientist who wasted no time in setting experiments up.

After months of delays, Baric was back where he belonged: surrounded by novel pathogenic coronaviruses. His first order of business was to probe the enigmatic cluster of bat coronaviruses, chief among them the SARS-like bat coronavirus RsSHC014, which had been unearthed by Shi Zhengli in the Shitou caves in Southern China. Under his watchful gaze, the virus—95 percent identical to SARS—had its spikes extracted and synthetically stitched onto a version of SARS that Baric described as a "replication-competent, mouse-adapted SARS-CoV backbone"—basically, a genetically engineered version of SARS that infected mice instead of humans. With the SHC014 spikes in place, Baric then added the chimera (bat coronavirus spikes on a mouse-adapted SARS strain) into a culture of human airway cells. He then injected the chimera into a handful of the hundreds of mice he had in his lab and infected other mice with a mouse-SARS hybrid strain he had produced. The mice were returned to their enclosures and placed back on a shelf.

Forty-eight hours later, Baric returned to the mice to see how his new creation had fared. It did not disappoint: the SHC014-mouse chimera, demonstrating a dazzling capacity for infection and evasion, had replicated just as well as SARS in human airway cells despite being adapted to mice; Baric probed further and found that the SHC014 spike was targeting the ACE2 ortholog gene, expressed by a section of DNA that humans shared with bats, mice, ferrets, and many other animals, making it an elegant shortcut into the cells of different species. SHC014 had the skeleton key that primed it perfectly for a zoonotic jump. Mimicking the moment when it did, Baric tried to use monoclonal antibodies taken from SARS patients to try to kill off the chimera; given that the two viruses were 95 percent identical, there was a chance that the blood of those who had survived an encounter with SARS might be well armed to destroy its younger cousin. If it was, then humanity had a hope against the most likely future coronavirus. All we would need to do was mass-produce antibodies and have them ready for when the next pathogen emerged.

But the chimera evaded the attack entirely; the antibodies, despite the similarities in the viruses, were useless. Following along the thread of a real-world epidemic, Baric moved on to the next defense humanity would eventually employ: a vaccine. When a SARS-like coronavirus emerged, Baric well knew, the road to creating a bespoke vaccine would be measured in years. But there was a strong chance that that virus would be as similar to SARS as SHC014 was, opening up the possibility that an off-the-shelf SARS vaccine, available on day one, might just work.

But when Baric inoculated mice with the SARS vaccine and then infected them with the SHC014 chimera, the off-the-shelf vaccine candidate didn't provide any protection against the bat coronavirus. But it wasn't just that the vaccine was ineffective: the inoculated mice, especially the older ones, actually got sicker after SHC014 infection than those that hadn't received the vaccine. The cure was a poison pill.

Baric emerged from the experiment clear-eyed and with a sense of dread. Humanity wasn't just underprepared for the next coronavirus. None of our scientific advances would make a shred of difference in the battle against the next pathogen the *Coronaviridae* family conceived. If anything, the existing arsenal of treatments and vaccines would probably make things worse.

It was a nightmare scenario. Baric had stitched the SHC014 spike protein onto the genetic backbone of SARS; this single act of recombination had created a deadly new pathogen capable of infecting human cells. SHC014 revealed to Baric that all the materials needed for a coronavirus pandemic—spikes capable of entering human cells, a genomic "backbone" that made it able to survive in human hosts, and tens of millions of recombining variants—were already circulating in the hidden virome. And if the pieces were all there, it was just a matter of time before they recombined.

No matter what it took for him and his team to get there, Baric's experiments on SHC014 were akin to moving the pandemic doomsday clock one minute closer to midnight. But beyond helping to show that nature had already procured all the components CoVs needed for spillover, Baric's acquisition of SHC014 was part of a grander plan. Once in his lab, the novel bat coronavirus was added to a panel he had been assembling since the 1990s—slowly at first, but with increasing speed as new coronavirus strains were collected, purified, and sent his way for analysis. By 2016, Baric had a collection that spanned the whole range of the *Coronaviridae*'s many lineages, subgroups, and variants. There were multiple variants of the six human CoVs (OC43, NL63, 229E, HKU1, SARS, and MERS) along with his decades-old laboratory tool, MHV. And then there were the new discoveries: RsSHC014, RaTG13, WIV1 (another "pandemic-ready" bat coronavirus that Shi Zhengli had discovered), and Baric's growing collection of bespoke synthetic recombinants, monstrous chimeras designed to probe the deadliness of emerging coronaviruses—in mice.

Baric, originally attracted to coronaviruses by the puzzle of their huge genomes, likened his collection to an encyclopedia ranging from volume one to fifty, encompassing the totality of the known members of the *Coronaviridae*. It was also, after SARS and MERS, his own personal early warning system. "Once the MERS outbreak occurred," Baric said, "it was like, 'Okay, we got lucky. We're lucky again. This is the second time we got lucky. We're not going to keep getting lucky.'" He was intent on using the viruses he had captured to make sure humanity was ready when its luck ran out.

Baric had long been collaborating with Mark Denison, the pediatric immunologist with whom he had discovered the secret behind the *Coronaviridae* family's gargantuan genomes. Denison, a soft-spoken scientist who often wore a wry smile, was of a similar cast to Baric. Both were just as fluent talking about specific genome sections as they were about how coronavirus epidemics moved through human society. Where they differed was in their orientation. Denison, a medical doctor, was keenly interested in understanding what about coronaviruses made them cause disease, whereas Baric was more naturally drawn to probing how their various structural components made them such elegantly efficient pathogens. Both, though, were confounded by the question of how to stop them. Nothing that modern science had developed to date worked, and the only reason that the world at large wasn't terrified was because the only two pathogenic coronaviruses appeared to burn themselves out before they spread too far. Baric and Denison had long since stopped looking back at the so-called victories against SARS and MERS; when they turned their gaze to the future, all they saw were countless coronaviruses testing our species. But how to stop them?

Like all good scientists, they went back to first principles. By 2015, Baric and Denison had been collaborating on and off for over twenty years. Up until that point, the highlight of their creative union had been their discovery of the coronavirus proofreading enzyme, that sneaky mechanism the family used to keep its offspring free from error during replication and which had allowed its genomes to balloon in size. Back in 2007, Baric and Denison had experimented with deleting the part of the coronavirus genome (nsp14-ExoN) that they thought encoded the proofreading enzyme; when they did, coronaviruses produced copies of themselves so riddled with errors that they ended up collapsing in on themselves. It was a neat trick, one that proved beyond a shadow of a doubt that they were right.

Over the ensuing years, Baric and Denison searched for ways to blunt the escalating threat the family posed. Each time they went hunting for weaknesses they could exploit, they kept coming back to their original discovery. The proofreading enzyme was embedded in every single member of the *Coronaviridae* and it hardly ever mutated across strains. Without it, the viruses immediately began to falter.

The question that hooked Baric on coronaviruses, and which took twenty-five years for him and Denison to answer, was how their genomes could be so impossibly large. After the emergence of SARS and MERS, the duo questioned how they could stop the inevitable coronavirus pathogen that would next emerge. The answer to the first question was the proofreader, nsp14-ExoN. But every single strain that the family produced required the proofreader to survive. And that meant that the answer to the second one had been right there in front of them the whole time.

If Baric and Denison could somehow cause the proofreader encoded in every coronavirus strain to fail, then any viral copies they tried to make would collapse into error catastrophe. Even if they didn't know what the next pathogenic coronavirus would look like, Baric and Denison knew that whatever emerged would also contain the proofreader. nsp14-ExoN was one among many such proteins located in the open reading frame, a vast region roughly twenty thousand nucleotides long that made up the bulk of coronavirus genomes. Without nsp14-ExoN, coronaviruses couldn't exist, which made the proofreader among the most defining features of the family. It was the viral equivalent of that famous Socratic syllogism: The *Coronaviridae* all possess a proofreader. All future coronaviruses will be a member of the *Coronaviridae,* therefore all future coronaviruses will possess a proofreader. And that, in Baric's eyes, made them mortal.

The key would be to find an antiviral drug that somehow blinded the proofreader, and could do so no matter what coronavirus it was targeting. "If you know that it works against volume one," Baric said, referring to his encyclopedia-of-coronaviruses analogy, "and volume fifty, and against all the volumes in between, you've got a pretty good bet that it would work against the unknown that you didn't test." With the panel and the plan, all he and Denison needed was the drug. Between the gain-of-function pause and the low threat level ascribed to SARS and MERS, coronaviruses were also a low priority for research funders. In short, there was no way Baric and Denison would be able to convince anyone to fund them to design a bespoke drug, which on average cost $2.6 billion. So, they had to get creative.

The two scientists drew up a wish list of attributes that a broad-based anticoronavirus drug needed to have: It would have to interrupt viral replication by introducing errors into the coronavirus genome. It had to be small enough to pass through cell walls without any fancy technology, and it had to be able to render the coronavirus proofreader useless. This led them to the large class of small-molecule drugs (so called because they can pass easily through cell walls) that had been developed as antiviral medications, many of which ended up languishing on dusty shelves, unused. Sometimes this was because they just didn't work against their intended viral target. Sometimes they were supplanted by more advanced therapies. And sometimes, the market for the virus they were designed to destroy had simply dried up. But where pharma saw useless intellectual property, Baric and Denison saw potential research gold. A massive library of simple antiviral drugs existed, all registered via the U.S. Patent Office, which numbered in the hundreds of thousands, and were ripe for the picking. Somewhere among them, there just might be a few that worked to disrupt coronavirus replication. The odds were low. Still, all they needed was one among the multitude.

Baric and Denison's long-standing scientific friendship was the central force powering their hunt for a pan-coronavirus antiviral. The intimacy they had developed over decades allowed their minds to swim together, their thoughts merging into one, the identity of the originator of any single idea lost amid the steady stream and winding tributaries of scientific advancement. Broadly, though, where Denison focused on how coronaviruses developed resistance to treatments, Baric's piece was the genetics, which meant drilling down to each coronavirus component to understand what it did and how it changed. He also brought his expertise in working with mouse models and unique human cell lines, as well as the concept of a universal pan-coronavirus therapeutic. The idea, he explained, was "not just trying to make one bug, one drug," by which he meant a drug that was designed to combat just one coronavirus among many. "With emerging pathogens, you don't know what's coming next, and you could have a great drug on a shelf that's collecting dust in the next coronavirus pandemic. That just seems shortsighted to me."

Baric and Denison began by narrowing down the most likely candidates from the vast archive of small-molecule drugs developed since 1899, when the first, aspirin, went on the market. (Nowadays, roughly 90 percent of the pharmaceutical market is made up of small-molecule drugs.) They zeroed in on nucleotide analogs, a type of antiviral that confounded their targets by swapping out nucleotide bases and replacing them with nonsense molecules that caused viruses to cease replicating.

It was arduous work. There were literally hundreds of thousands of nucleotide analogs, and Baric and Denison had to painstakingly whittle down the candidates to the most promising ones based on their knowledge of viral activity, the known impacts of the drugs on other viruses, and their own experiments. They knew that it could take decades to sift through the candidates and find even a handful of drugs that worked against coronaviruses. But they had no choice.

They would spend years doing steady, plodding appraisals of their antiviral drug archives. This meant systematically experimenting with the most promising candidates by introducing them into a variety of human tissue cultures infected with multiple coronaviruses to see how they would fare. While most scientists testing viruses and drugs in vitro did so using just one type of human tissue (usually a single type of epithelial cell, the cell that lines the outer surfaces of blood vessels and organs), Baric didn't think this was good enough. Antiviral drugs might affect coronaviruses differently depending on which cells they were in, he reasoned. So, he set up a system of nasal airway epithelial cells, large bronchial airway epithelial cells, and multiple cells found in the lungs, which were all the places coronaviruses proliferated; then he infected the cells with coronaviruses and systematically added candidate antivirals one by one to see how well they performed in the various battlefields.

Denison, more naturally extroverted than Baric, had been actively seeking out promising candidates rather than relying on the drugs that he and Baric had pegged as most likely to be effective at dismantling coronaviruses. This had led him into ongoing talks with multiple pharma companies, big and small, to gain access to potential winners. He had some contacts at the biopharmaceutical giant Gilead, and shortly after he and Baric launched their plan in 2013, he approached them to see if they might be interested in having their shelved drugs tested against coronaviruses.

Denison knew that Gilead had accumulated a massive library of the patented antivirals it had produced with varying degrees of success. So, he offered them something for nothing: all the company would have to do was sign a waiver allowing him and Baric to test its drugs, and the virologists would do the rest. Seeing little market upside in coronaviruses—SARS was gone, and MERS had killed only a few hundred people worldwide—Gilead nevertheless saw no downside to Denison and Baric's request. If the drugs it shared didn't perform against coronaviruses, it was no big loss. If they did, it was another feather in Gilead's cap and a potential source of revenue, albeit limited. It agreed to hand off about thirty antivirals, which on balance were more likely to be effective than the vast archive Baric and Denison were testing. Baric expressed some surprise that the pharma giant had even agreed to engage with their future-facing project. In his view, despite the potential upside, there was still a long way to go from their abstract venture to the market's bottom line.

Gilead's thirty antivirals were at best a mixed bag. At worst, they were some of the least effective drugs the company had recently developed. The top candidate was an antiviral dubbed GS-5734 that had a reputation as something of a loser. Gilead had originally developed it in 2009 as a therapy for hepatitis C, but tests quickly showed that it didn't really work to prevent infection. The pharma giant, opportunistically seeking a home for its now-orphaned drug, repackaged it as an anti-Ebola agent during the 2013 West African Ebola epidemic; again, GS-5734 failed to make much of a dent.

After two major losses, the drug was shelved indefinitely—until Denison offered a lifeline.

GS-5734 was on a familiar path: just another promising drug that didn't deliver. But under the care of Baric and Denison, the drug's potential soon revealed itself. Unlike with hepatitis C and Ebola, GS-5734 performed spectacularly against coronaviruses, including SARS, MERS, and the two bat coronaviruses, SHC014 and WIV1, that Shi Zhengli had harvested from Yunnan Province. By 2016, Baric and Denison had discovered that GS-5734, almost unique among the hundreds of thousands of drugs they were screening, was protective against disease progression caused by coronavirus infection. Better yet, when monkeys were given the drug prior to exposure to coronaviruses, it was wildly effective at preventing them from getting infected. Baric and Denison had, after years of testing, seen a signal.

GS-5734—trade name remdesivir—was promising, but the virologists didn't stop there. In 2015, roughly two years after Baric and Denison began their quest, they were approached with a proposal. George Painter, the head of the Emory Institute for Drug Development, which had created a promising antiviral drug dubbed EIDD-2801, was looking for ways to test its capabilities. Painter, a giant in the field, had also been asked to sit on an external NIH board to review the data that Baric and Denison's project accumulated. He had seen the early results from GS-5734/remdesivir, which clearly demonstrated that the shelved antiviral was actually a pan-coronavirus assassin. Painter, who had spent his career testing drugs, appreciated Baric and Denison's vision of equipping humanity for a future war. He was also deeply impressed by the system Baric had set up to test how potential pan-coronavirus cures reacted in different cell types. When Painter's group developed EIDD-2801, and their early tests suggested it might have a potent antiviral effect, he had already been reading detailed reports about Baric and Denison's accomplishments for two years.

Painter approached Baric and Denison to see whether they'd include his new drug in their panel. Intrigued, Baric and Denison began putting the drug through the series of challenges and trials they ran for their most promising candidates. To their delight, they found that EIDD-2801, like remdesivir, could cause coronaviruses to become engulfed in error catastrophe, which ended replication. What was more, the drug had been formulated as an oral medication, meaning it could be distributed easily and taken by anyone; remdesivir, by contrast, had to be delivered by injection from a medical professional. Regardless of how they were delivered, remdesivir and EIDD-2801 gave Baric and Denison two options, albeit still only theoretically, to fight a future coronavirus pathogen.

Both drugs worked by swapping faulty molecules into the virus's RNA replicase, the section of the genome that acts as its copying machine. But neither operated with brute force. If the drugs had caused a ruckus, the coronavirus proofreader enzyme, like a truculent beat cop, would have bounced them immediately. Instead, both GS-5734/remdesivir and EIDD-2801 functioned with the sleight of hand of an experienced pickpocket. After subtly embedding themselves into the viral copying machine, they allowed the machine to keep on churning out copies for a few cycles before they began to feed errors into the genome. This pause confounded the proofreader, which then allowed the drugs to riddle the viral copying machine with abundant, overwhelming errors. Their action wasn't like a violent mugging on a busy street in broad daylight. Instead, it was akin to that sudden realization, after you've spent hours crisscrossing the city, that somebody, somewhere, stole your wallet.

The only hitch was that, in 2016, Baric and Denison ran into a problem all too familiar for scientists studying the family. They had discovered that remdesivir and EIDD-2801 were effective in mice, monkeys, and human cells in a petri dish. But there were too few actual humans infected with pathogenic coronaviruses to confirm that the cures worked in the real world. All they could do, after years of exhaustive discovery, was wait for a future threat to prove them right—or wrong.

BARIC SAW CLEARLY THAT THE GREATEST THREATS LAY BEFORE US, NOT behind. Though the time, place, and strain were beyond his capacity to predict, he nevertheless knew that humanity's battle with coronaviruses was not yet over. In lectures, he started jokingly referring to himself as the fifth horseman of the apocalypse, a figure heralding a coming doom. Meanwhile, in his lab, his panel displayed before him, there was always so much more he could do to save as many future lives as possible, if only he kept on probing the mysteries that the *Coronaviridae* still hid from view. And while he worked, deep within the virome, the virions collided, shifted, and recombined. The churning never ceased, new strains and variants emerging in every conceivable orientation, a bottomless pit of protean possibility endlessly testing the defenses of humanity barely beyond its reach.

Chapter 8

THIS IS A FINITE NUMBER

2008, Department of Stem Cell and Regenerative Biology,
Harvard University

BARIC AND DENISON'S CO-DISCOVERY OF THE CORONAVIRUS PROOF-reader was, at the time, a personal milestone for both of them. Baric had long hungered to understand the confounding viral family, driven by nothing more than an abstract desire for discovery; he wanted to map the unknown, and he had. Little could he have known, with remdesivir and EIDD-2801 ready and waiting, how consequential to human survival his speculative journey would become. But he was far from the only scientist whose arcane scientific pursuits would end up illuminating a path through the fury and carnage of the *Coronaviridae*.

Derrick Rossi, a peripatetic stem cell biologist, had a roving mind and an appetite for discovery that appeared to know no bounds. A self-styled iconoclast, he carried himself with a brashness diametrically opposed to the typical conception of a basic scientist. Despite being in his fifties, he had a boyish look about him, and he was resolutely committed to his soul patch, thick-rimmed glasses that recalled Elton John's more fabulous looks, and band T-shirts professing his love of punk rock. Quick to laugh and prone to gesticulating wildly when he spoke, Rossi had an earnest, searching look when framing his thoughts; when he was on the subject of stem cells and the building blocks of life, those thoughts seemed to come too quickly, making it hard for him to stop and catch his breath. After stints in Toronto (Ph.D. attempt number one), Paris (Ph.D. attempt number two: "I couldn't finish that because of exhaustion"), Dallas ("of all places, Dallas, Texas"), Helsinki (Ph.D. attempt number three: "I've never been to Finland, so I'll go live in Finland"), and Stanford (a four-year postdoctoral fellowship), Rossi landed at Harvard in 2008, where, true to form, he decided to switch tack yet again. Though his Ph.D. was in molecular biology, he had a new obsession: stem cells. "I was very interested in aging," he explained, "and specifically in stem cell aging, because it appeared to me that aging might be underwritten by tissue-specific exhaustion, basically, that accumulated over the years." Having dreamed up this hypothesis, he became single-minded in exploring it.

Throughout his many attempts at a Ph.D., Rossi had developed an almost unrivaled skill set in genetic science and a track record of publications to prove it. To hear him tell it, knocking out genes in mice—which amounted to taking sections of naturally occurring DNA, replacing them with sections designed in the lab, and then breeding more mice using the altered genetic code—became second nature to him. Genomes were just complex puzzles to unlock, explore, rearrange, and make better—which he did through sheer will and electrifying skills. In the lab, his exceptional tactile prowess and fearless creativity allowed him to push experiments far beyond what his supervisors and colleagues had ever intended. Rossi hated being buttonholed into mapping out exactly how an experiment was going to end. What, after all, was the point of doing science? Why bother doing experiments if you weren't going to let them evolve? So, even as his reputation for excellence grew, he chafed at what he saw as the unnecessary constraints of the academic scientific process.

Still, Rossi carved out his own home as a junior faculty member at Harvard Medical School, where he was free to pursue his research on the mechanics of stem cell aging with little oversight. An oblique thinker, he was always running side projects in his lab, research tributaries that led to unknown places, often inspired by the work of other scientists he consumed voraciously. In 2006, he was reading a paper by Shinya Yamanaka, a Japanese stem cell researcher who had discovered a method to coax mature cells to revert into stem cells—in essence, to reverse the process of biological aging. It was groundbreaking work, with implications across the biomedical spectrum (most immediately, as a way to treat cancer cells), for which Yamanaka would later win the Nobel Prize. Rossi was blown away; it was breathtaking and beautiful science. Yamanaka had managed the feat by delivering instructions into cellular DNA using a virus—but not just any virus: Yamanaka used retroviruses, which insert themselves into the genome of their hosts after they infect them. Once enmeshed in their hosts' DNA, retroviruses fully integrate with it and, as a result, are next to impossible to dislodge. (This integration was a key reason that HIV, a retrovirus, has been so hard to vaccinate against.) For Yamanaka, this genomic fusion was an excellent way to send commands to a cell. But his technique had a weakness. Once enmeshed, his retroviruses were inadvertently making cellular DNA express proto-oncogenes, precursors to cancer, which could be very dangerous. Rossi figured he could do better.

Rossi had a favorite saying: "DNA makes mRNA makes protein makes life." It was a shorthand he used to explain, in the crudest of terms, the complex pathways by which genomes produced messenger RNA (short fragments of genomic instructions), which then caused cells to express specific proteins (the macromolecules that make up the basic building blocks of all life). With that simple chain of events in mind, Rossi figured that if the goal was to instruct cells to reverse the aging process, the easiest way to do that would be to follow the body's own processes. "If we don't want to integrate viruses into the DNA, you could either send instructions with mRNA, or you could do it with proteins." Proteins were, Rossi said, bulky and inefficient tools for directing cellular activity, which left him only one option: mRNA.

The problem was that no self-respecting molecular biologist used mRNA. The chief reason was ribonuclease, an enzyme found through-out organisms that digests and utterly destroys any RNA fragments it comes across. Ribonuclease was so ubiquitous that, in most labs, it was difficult to find a surface clean enough to experiment with RNA; often, as Bob Brunham and Ian Lipkin well knew, RNA would simply degrade upon contact, ending the experiment before it even started. Even if the mRNA got into a cell, ribonuclease would destroy it immediately. But the entire field's having moved on from mRNA didn't convince Rossi that he should, too. His self-styled punk rock persona had imbued in him a knack for going against the grain.

Rossi and his lab partners figured they'd at least try to improve upon Yamanaka's delivery system by using mRNA rather than a virus to reprogram a cell's DNA. They synthesized a short, standard mRNA—"just beautiful-looking mRNA"—and inserted it into a cell, taking care to keep the space as clean as possible to avoid the inevitable ribonuclease attack. Their experiment failed miserably. Not only was the cellular DNA unchanged, but the cells that came into contact with the mRNA died. Rossi, it turned out, had inadver-tently triggered an ancient antiviral self-destruct sequence. As soon as his mRNA molecules passed through a cell's wall, the cell de-tected them as foreign substances and responded by shutting itself down in a process known as altruistic suicide, a last-ditch effort the immune system deploys to stop a viral invasion. Rossi was relearn-ing what many molecular scientists had known since mRNA was first discovered in the 1960s: that the substance was, despite its po-tential applications, nothing more than a very good way to wreck cell cultures. "There's other ways to kill cells in the dish," Rossi said, with obvious sarcasm. "Just put them in the microwave: it's easier."

Rossi was convinced that if he could just solve the problem of the delivery, the promise of using mRNA as a therapeutic tool would be boundless. So, he did what came naturally: he returned to the scien-tific literature, obsessively consuming write-ups of experiments that might hold the clues to his mRNA problem. He couldn't be the only person in the entire world who had faced this problem, he figured.

He was right. Two virologists at the University of Pennsylvania, Katalin Karikó and Drew Weissman, had been tinkering with viral RNA to understand how it triggered the human body's antiviral response—the same response that was leading Rossi's cells to commit altruistic suicide. In 2005, after testing how cells reacted to HIV and herpes RNA, Karikó and Weissman discovered that if they swapped out a few key nucleosides on the RNA (the individual bases that make up the script of the genetic code), it would no longer cause a cell to destroy itself. Karikó and Weissman had discovered a kind of molecular camouflage that let RNA move through cells without triggering a scorched-earth response.

Rossi was rapt. Though Karikó and Weissman had tested only viral RNA in their experiment, he saw no reason not to apply the same concept to his synthetic mRNA. Following the duo's recipe, Rossi swapped out the key nucleosides in synthetic mRNA and then programmed in a gene that expressed fluorescent jellyfish protein, which was written into the DNA of the ghostly sea creatures and gave them their uncanny green glow. Rossi added the modified mRNA to a container of cellular tissue and then sat back and watched. Soon, instead of a plate full of suicidal cells, before him lay "very happy, healthy, green cells," a literally glowing testament to the success of his experiment. It was, Rossi said, a true eureka moment. He had found a way to introduce instructions into a genome without altering its DNA, without destroying the messenger, and without killing the cell.

Rossi quickly ran through other experiments to test the bounds of his discovery. These included creating a synthetic mRNA with instructions to have a cell produce luciferase, the protein that causes fireflies to glow, which he injected into the thighs of lab mice. He anesthetized the animals and placed them in a dark chamber, then set up a photosensitive camera to watch. Soon, small orbs of light began to shine, pulsing through the thin muscle and skin of the mice's legs. The technology felt limitless. "Whenever we went to express a new protein, it was always, you know, just easy," he said. It would take him and his team about a week to build a DNA construct from which he would synthesize new mRNA instructions, which always included the modified nucleosides from Karikó and Weissman's recipe. Rossi used the modified RNA technique so often he even came up with a shorthand to describe it: mod-RNA. When, in 2010, the moment came when he wanted to commercialize his discovery, he had already settled on a company name: "Moderna."

In his opening pitch to investors, Rossi ran through the many applications he envisioned for his technology, including better diagnostic tests, therapeutic drugs, and gene-based therapies. There was one application, though, that had completely eluded his unbounded imagination: vaccines. He freely admitted that the notion hadn't even crossed his mind. "I wish I could say that I had that insight," he said, "but I didn't. And for the record, nobody at Moderna did, either. At least in the first three or four years, there was no talk about a vaccine." Rossi was blunt about his rationale. "It was," he said, "a nonviable business model."

BY 2015, THE U.S. GOVERNMENT'S VACCINE RESEARCH CENTER WAS at something of a crossroads. Housed in Bethesda, Maryland, at the National Institute of Allergy and Infectious Diseases, the Vaccine Research Center, or VRC as it was called, had been laser-focused on the elusive goal of an HIV vaccine for more than three decades and counting. The deputy director of the VRC was Barney Graham, a towering and soft-spoken clinical trials physician who, in 1982, was among the first to treat AIDS patients in the American South, when HIV was still a largely unknown pathogen. Graham, who had retained a genteel manner from his childhood in Kansas, also exuded an immediate and overwhelming empathy. In conversations, he gave the sense of embodying his interlocutors so as to better understand their point of view and to gently provide them with all that they needed from him. This had served him well throughout his career with patients and colleagues alike.

When, in 1988, Graham was tasked with leading one of the first American clinical trials to test an HIV vaccine, the initial results were overwhelmingly promising. The patients given the vaccine produced a strong antibody response, which signaled that the body was arming itself to kill the virus. Graham was ecstatic. "It was a time of hopefulness," he said. "But we were so naïve." His optimism abruptly faded when the vaccine candidate he was trialing, despite clearly eliciting antibodies, did nothing to stop people from becoming infected with HIV and eventually dying. When all the numerous vaccine candidates in that first round of trials failed, the first seeds of doubt started to emerge that HIV, which all had predicted would be cured within two years, would be a harder target to control. "There was a lot of excitement then," Graham said, "but it eventually just turned into frustration."

The puzzle of HIV forced vaccinology into retreat. Before the virus emerged, vaccine scientists like Graham generally didn't know why their vaccines worked; nor did they really care. The field was broadly aware that the immune system could produce antibodies that protected the body from a virus. But beyond that, there had been little impetus for anyone to explore how that process functioned. When the vaccine platforms didn't work against HIV, though, vaccinologists were forced into a crash course in human immunity.

Facing hurdle after hurdle in outwitting HIV, Graham had, since the 2000s, been exploring circuitous paths to find answers to the vexing problems the virus presented. As he went down research rabbit holes, he began to build a grander vision, one that went far beyond HIV or any other single pathogen. Over the past fifty years, he discovered, the number of new viruses infecting humans had been steadily increasing. Counter to that trend, though, the number of viral families that included human pathogenic viruses had plateaued. This disparity—between the accelerating onslaught of individual viruses threatening humans and the static number of viral families from which they emerged—provided Graham with the broad contours of the problem. "The point was that this is a finite number. This is a big but tractable, do-able thing."

That "thing" was creating at least one vaccine that worked against a member of each of the twenty-six viral families with the ability to infect humans. He knew it sounded audacious, but he also knew that having a representative vaccine on the shelf for each family, even if it wasn't exactly tuned to the specific pathogenic strains that might emerge, would dramatically accelerate the development of effective vaccines against emergent pathogens whenever they might arise. When the next viral threat presented itself, he figured, all that would have to be done was some tweaking. So, he got to work.

Graham's decades in vaccine development and his generation's failure to produce an HIV vaccine had taught him a key lesson: architecture mattered. Though a dumb collection of molecules, amino acids, and genetic information, viruses aren't static. A virion's brief existence—its life cycle, for those willing to go that far—is a frenzied, fitful process of separation, binding, fusion, unfurling, and replication. Throughout, the virus is interacting with its host, alternately attacking and reacting. As each virion goes through its motions, its various components, which are initially latched tightly in place, are unlocked one by one, springing open like an elaborate Swiss clock. Each of those shapeshifts alters the virion's configuration to fit the specific need of the moment, be it binding to a cell wall, fusing with it, or using the current of its host's life forces to float into the nucleus, seize the machinery, and begin to replicate.

In 2008, when Graham and the vaccinologist class had already been toiling toward the ever-elusive goal of an HIV cure for twenty-five years, advances in structural biology allowed them finally to trace the metamorphic architecture of viruses. Using X-ray crystallography, a technique that created three-dimensional images of molecular structures, Graham and his then postdoc Jason McLellan charted topographical maps of viruses in each of their changeable states. Starting with a virus called RSV (respiratory syncytial virus), which caused mostly minor colds in adults, Graham and McLellan found that RSV's F protein (equivalent to the coronavirus spike) radically changed shape after it fused with human cells, morphing from the shape of a bouquet of daisies into a spent matchstick.

Tracking this metamorphosis wasn't just a scientific curiosity. In 1966, scientists had developed an RSV vaccine specifically targeted to children and infants, among whom infection was more severe and the risk of hospitalization high. With few technologies available, making a vaccine in the 1960s was pretty straightforward. In the case of the RSV vaccine, the virus was extracted from a culture, put in formaldehyde to kill it, and then injected into infant test subjects. What happened next confounded scientists and led to one of the more disturbing episodes in the modern era of vaccine making. Within a few days, the inoculated infants had developed antibodies against RSV, just as the scientists running the trial had hoped. But then the world seemed to tilt sideways. Despite the antibodies, the vaccinated babies got sick. And then they got sicker: 80 percent of the infants given the vaccine ended up in the hospital with RSV, compared with only 5 percent of those who weren't vaccinated. Ultimately, two of the vaccinated babies died. Work on an RSV vaccine immediately ceased.

It was horrific. It was also a mystery: Why, if they generated robust antibodies, did the babies die? It would take until 2016 for Graham and McLellan to answer this question. "We now know that the way that they made that vaccine completely flipped all the F proteins from the pre-fusion to the post-fusion state." This shape change, innocuous though it might seem, meant that when the infant immune systems were presented with the vaccine containing the killed virus, they created antibodies expertly designed to target the shape of RSV virions after they had fused with a cell. This misdirection rendered the antibodies useless when the infants became infected, because the virus entering their systems had a totally different shape from the one they were inoculated against. Worse, Graham explained, the immune cells that the infants produced ended up getting lodged in their tiny airways and, in his words, "caused a lot of destruction and disease." The vaccine itself was toxic.

It was a sobering lesson, one learned fifty years too late. But Graham, having failed to produce an HIV vaccine for over three decades, knew better than anyone that no scientific knowledge, especially that born of failure and death, should ever be dismissed as useless. He became convinced that, no matter the viral pathogen a vaccine was programmed to protect against, the shape of the vaccine's payload was essential to how effective it would be.

Beginning in 2010, the Vaccine Research Center expanded in all directions to test this hypothesis, while also kick-starting Graham's quest to create a vaccine for each of the viral families known to infect humans. For each vaccine candidate, Graham and McLellan artificially manipulated the specific viral antigen (the part of the virus that antibodies attack) delivered by the vaccine. This involved using prolines (bundles of about twenty nitrogen, hydrogen, carbon, and oxygen molecules) to pinion the antigens in their coiled, prefusion state. In each trial, the vaccine candidates they had designed using these engineered proteins acted like glowing beacons for human immune systems, causing them to produce a rush of antibodies perfectly fit to conform to the shape of viruses and then to kill them.

When MERS emerged in 2012, killing the Bisha index patient and spreading death across the Gulf states and beyond, Graham became alarmed. While it moved slowly through human populations, it was another new pathogenic virus among the many that had recently emerged, a trend that looked more and more like an exponential acceleration. Along with MERS, there was Chikungunya (2004), Zika (2007), and the West African Ebola epidemic (2014), all of which were essentially uncontrolled. "It just kept going on and on," Graham said, as new viruses spilled over into humans, "and we wanted to try to exploit these technologies to get ahead of pandemic threats."

Back in 2003, when his group tried to develop a SARS vaccine, it didn't work, but he didn't have the technology then to understand why. In 2013, when Graham probed the architecture of the MERS virus, he found that, just like RSV's F protein, the receptor-binding domain that capped the spike protein was, basically, a microscopic spring-loaded mechanism that popped open and unspooled itself after it came in contact with cell walls. At that point, the shape of the spike protein changed radically. But, Graham figured, if he could prevent the spring-loaded mechanism from popping open, then the immune system would generate antibodies that fit the shape of the virus when it first entered the human body rather than after it had already penetrated cells, by which point all was already lost. Graham's team perfected an artificial MERS spike protein in their lab that they pinioned in place to hold its pre-fusion state. With that payload in hand, Graham moved to the next phase, which he dubbed a prototype pathogen demonstration.

Graham had the MERS spike protein, but he needed a delivery system to get it into human bodies. As he surveyed the landscape of opportunities in 2015, one thing was clear: the same old approaches wouldn't cut it. Luckily, he knew just where to look.

March 2015, Cambridge, Massachusetts

At Moderna, the mRNA technology pioneered by Derrick Rossi promised many things. It was a way to build diagnostic tests. It was a perfect approach to target cancer cells. And it was a delivery system for gene therapies designed to treat rare diseases. But by 2015, at which point Rossi had left the company he founded, some scientists at Moderna had begun to believe that the company could be something else: a vaccine maker, something that hadn't ever crossed Rossi's mind. Tal Zaks, who joined as chief medical officer in March 2015, after years in the executive suite at the pharma giant Sanofi, was among the vaccine evangelists.

Zaks, a clinician-scientist, had spent practically his entire career straddling pharma and academia, and exuded a fervent belief in the industry's ability to solve the most vexing problems posed by nature. In his early fifties, Zaks was at once youthful and patrician, the shock of wavy hair extending forward above his oft-knitted brow giving him an adolescent air. He spoke at a rapid clip, his speech betraying little of his upbringing in Israel, and as he delved deeper and deeper into the technical, he hunched under the weight of his ideas like a linebacker readying himself for a hit. His hands, which constantly gestured, did much of the talking, and Zaks had a tendency to raise them higher aloft the more he became consumed by his ideas.

When Zaks arrived at Moderna, all the applications that the company foresaw for its mRNA tech relied on the same basic mechanism of action. The company's synthetic, preprogrammed mRNA strands would deliver encoded instructions into human cells, where they would then instruct DNA to make proteins that were clinically useful for patients. The problem that company scientists had been running into was that human cells didn't appreciate being invaded by foreign objects and would mount an immune response to try to stop the mRNA from trespassing. The big fear was that Moderna's mRNA-delivered gene therapies, which instructed cells to produce proteins that some people weren't otherwise able to make, would inadvertently cause the body to target the very protein that could cure their disease. "You would cause immunity against the protein that you're making," Zaks explained. "If you're making a protein to replace a missing gene in a rare disease; well, in a kid with a rare disease, that's not a good thing." This led to a choice: solve the immunity problem or lean into it. "And so," said Zaks, "a few people at Moderna asked, 'Well, how can we take a bug and turn it into a feature?'" For a young biotech company looking to burnish its credibility, Zaks believed, a vaccine was the simplest way to prove to the world the power of its futuristic technology. For all its other applications, Moderna had to show that its mRNA made it into a cell and instructed the cell to make a protein and that the new protein had some kind of therapeutic benefit. This was a complex and hard-to-measure set of actions. With a vaccine, Moderna was able to take away that last step entirely. As soon as the mRNA taught a cell how to make a viral protein (say, a coronavirus spike), the body would immediately react to it. All Moderna had to do was measure the antibodies that were then produced. As Zaks put it, "Vaccines are the low-hanging fruit. You need a smidgen of protein, and the immune system does the rest." Compared to treating rare diseases like cystic fibrosis, which required huge amounts of mRNA-delivered proteins to be produced in exactly the right place, vaccines were easy.

By 2015, despite the potential Zaks saw in the technology, Moderna's vaccine program hadn't really gotten off the ground. He was intent on changing that. After he was hired, the new chief medical officer's first all-company presentation, titled "How Moderna Will Save the World," featured images of the Spanish flu and unclassified reports from the U.S. Department of Defense warning that a future pandemic would lead to the deaths of as many as two million Americans. In slide after slide, Zaks hit home on the advantages that an mRNA vaccine would have over traditional platforms, including rapid production and a capacity to reprogram mRNA strands to attack new variants.

Between 2015 and 2019, Zaks and others worked on developing an in-house vaccine program focused on potential pandemic-level viral threats. This included testing vaccine candidates for two newly emergent strains of avian flu and for Chikungunya on animals, including primates, and then one strain of avian flu on humans. Zaks loved it. As a physician, he was single-minded about getting scientific innovations out to real patients. Even better, the vaccine candidates were showing clear signs of working.

Moderna's entry into the vaccine space came just as the world was accelerating into a period of emerging viral threats. Since 2000, at least eight novel viruses (including SARS, MERS, Ebola, four avian flu strains, and Chikungunya) had emerged, all of them with pandemic potential. With the rising zoonotic tide as a backdrop, the success of Moderna's vaccine work naturally caught the attention of Anthony Fauci and Barney Graham at the National Institute of Allergy and Infectious Diseases, which in 2015 proposed collaborating with Moderna on a vaccine for Zika, a mosquito-borne flavivirus that caused a deadly fever and, among pregnant women who were infected, birth defects in babies. Zaks was elated to have finally entered into a partnership with the U.S. government. It was a key piece of credibility making that would undoubtedly help persuade the world that Moderna's mRNA technology, which for most people was an abstract moon shot, was for real. Within a year of the collaboration, the pharma giant AstraZeneca had doubled down on an early investment it made in Moderna; a host of venture capitalists and funders, including the Bill and Melinda Gates Foundation, followed suit. Though Zika would eventually fade away before Moderna and the NIH could trial an mRNA vaccine, Zaks foresaw great things. "The relationship with the NIH," he said, "was strong." It was also just beginning.

UNTIL INTEREST IN USING MRNA BEGAN IN THE 2010s, THE WORLD had relied on three traditional platforms that produced safe and effective vaccines. The first used "inactivated viruses," which involves isolating a virus, removing one of its key components to stop it from being able to replicate, and then injecting it into a host to cause an immune response without subjecting the body to an actual "live" virus. The advantages of using an inactivated virus platform are, chiefly, that there is no risk of the virus's replicating; those strange protein-encased virions simply bounce around the immune system like floating cadavers until they attract newly created neutralizing antibodies that, like hungry schools of sharks, devour them. But just like cadavers, the inactivated virus is less appetizing to the sharks than live flesh and might draw fewer of them; in viral terms, this means that the body's immune response to inactivated viruses is often much less intense compared to attacks from live viruses. The upshot is that the immunity conferred by inactivated virus vaccines is often partial, short-lived, and narrow (meaning inactivated vaccines are often effective against only a specific viral strain). One good example is flu vaccines, some of which use inactivated influenza viruses to jolt the body into producing antibodies but often aren't effective against more than a few strains. This creates that all-too-familiar situation: the annual requirement, that always seems to come too soon, for another booster; another needle in the arm—all to keep the body's immunity going.

The second traditional platform used what's known as an "attenuated" (or weakened) virus to prepare the body to fight off the stronger, wild version of the pathogen. An attenuated virus is still living, or as close to living as a virus can get, but it's been systematically engineered to the point where it can no longer effectively replicate once it's introduced into its host. To do this, vaccinologists pass human viruses through animal cell cultures in labs, taking advantage of the viruses' natural tendency to adapt to new circumstances. The virions that can adapt fast enough in the animal cells will multiply repeatedly, ultimately replacing the original version of the virus, while those virions that don't adapt, too ardently bound to human cells, die out. Ultimately, as the generations of virions live, die, adapt, and replicate, the whole saga captured in the edgeless boundary of a petri dish in the space of a few weeks, the new version of the virus, so well adapted to the animal cells within which it has grown, will emerge as a frailer version of its ancestor. Only then is it introduced into human hosts, where it can no longer replicate, though it will still generate an immune response. As with the clever fish that evolved to walk and breathe air, the virus's evolution makes it difficult for it to survive and thrive in the environment from which it first emerged; once you grow lungs, you can drown. But there is a potential wrinkle: unlike inactivated viruses, attenuated viruses still in principle have the tools to replicate inside a human host, even if they've forgotten how to use them. While the risk that they will mutate back to their original form is exceedingly rare, it can still prove deadly.

Dead viruses, attenuated viruses—if the one was weak, the other might just be the cure that ended up making you sick. The third viral vaccine platform, known as a subunit vaccine, sought to sidestep both these problems by introducing a small piece of the virus into the body instead of the entire pathogen. All viruses, despite their simplicity, contain multiple elements: a genome sheltered by a structure of molecules, along with proteins that allow them to keep their shape within the rigors of their host's biological environment. But that's not enough: viruses need also to penetrate cells to continue the cycle of replication, which they do through proteins on their surface that serve, like battering rams, as weapons of cell invasion; the coronavirus spike is exemplary of how effective a weapon these proteins can be. Subunit vaccines extract these weapons from the rest of the invading force—a battering ram with no one to swing it—and introduce them into our bodies like relics behind museum glass, inert weapons of brute destruction. This alerts the immune system, which then produces antibodies specifically targeting the molecular weapons in anticipation of a real attack, without placing the body at risk of an actual invasion. In this way, subunit vaccines are both very safe, because they contain only a nonreplicating portion of the virus, and generally quite effective, because they specifically target the tip of the spear of viral attacks. The one drawback is that the body's response to subunit vaccines can be middling or weak. To counter this, subunit vaccines typically include adjuvants, often chemical salts, that enhance the immune response.

While all these vaccine platforms had proven safety records, Zaks split them into two camps. "The field had diverged a few decades ago," he said. "One is the old, inactivated or attenuated viruses, which people could fiddle around [with] and design easily. But a whole virus is immunogenic in a way that you don't fully understand, so you're just putting it in and hoping that it would work." Zaks respected anything that helped save lives; still, he was evidently dismissive of the science behind killed and inactivated viruses. It was simply inelegant to rely on a technology you couldn't comprehend. "The other part of the science said, 'No, no, we actually understand what the immunogen ought to be, and we will direct the immune system against that protein.'" Zaks was referring to the subunit vaccines that isolated the most important parts of a virus (say, the coronavirus's spike) to make sure the immune response was as targeted as possible. But those weren't always enough, either. "A protein in and of itself when you inject it is just too innocuous. You need to somehow amplify the signal. You need to tell the immune system somehow, 'Hey, there's something bad there. Go pay attention.'" That's why vaccine makers used adjuvants to enhance the immune reaction, though doing so had its drawbacks. Adjuvants could be toxic, especially for children, which forced vaccinologists to find a balance between creating vaccines powerful enough to protect people but weak enough not to poison them.

In the case of coronaviruses, there was strong evidence that human bodies just weren't very good at repelling them, which is why the colds caused by OC43, NL63, and 229E were so common. We don't stay fully immune beyond a few months and are doomed to catch them over and over again.

While the number of potential vaccine platforms is in theory limitless, there were three experimental ones that, in 2015, had yet to deliver on their promise. The first used the shell of an adenovirus (a family of viruses that causes most common colds) as a way to deliver instructions to the body on how to kill SARS-2. Adenoviruses are useful vectors because their shells are much more rigid than that of other viruses, making them able to withstand more of the firepower our immune system uses to attack them. Once an adenovirus vaccine "infected" a human host, the theory went, it would quickly make its way into our cells, release its genetic material, and instruct those cells to start making copies of themselves, which would in turn alert our immune system to launch a counterattack, which it does very well against adenoviruses. This potent virus–host interaction could be exploited by removing the adenovirus genome and replacing it with parts of another virus's genome. The assumption, largely unproven (though at the time, being explored by two large pharmaceutical companies, AstraZeneca and Johnson & Johnson), was that an adenovirus-delivered vaccine would then cause the body to mount a powerful defense. This would be elicited by the adenovirus shell, but the immune system's mustered arsenal of antibodies would instead be directed to attack the viral material contained inside it.

The second experimental platform, DNA vaccines, dispensed with the virus altogether. DNA vaccines could, in theory, use the basic building block of life to take over cells and instruct them to produce a specific type of viral protein against which the body would then launch an immune response. The benefit of using DNA is its resiliency: its massive double-helix structure keeps it protected from the bruising environments it encounters within a human body. This size and resilience present a logistical challenge, though. "You would think that a DNA vaccine may be a little bit more stable and may have advantages, but in fact the opposite is true," said Zaks. The reason is basic physics. How do you get something as big as DNA through a cell wall? In the molecular universe, DNA strands are juggernauts, long, twisted affairs containing tens of thousands of paired nucleotide bases organized into a complex and tightly bound architecture, their sheer size dwarfing the other structures that surround them, like battleships looming over inflatable dinghies. This bulky, two-strand coil of base pairs is too hefty to easily slip into cells. Without a way to pry their walls open, a DNA vaccine would be dead in the water.

And then there was mRNA, the wispy strand of nucleic acids long believed unusable given how prone it was to collapse at the lightest touch. But where many saw mRNA as woefully ill-suited to the rigors of intracellular travel required of vaccines, Zaks saw something else: the end result of billions of years of evolution. "What's the most abundant class of viruses?" he asked. "RNA viruses—and that's for a reason." To hear Zaks tell it, Moderna's mRNA vaccine platform was adopting the tactics used by the viruses it was designed to destroy. RNA viruses (a class that includes the *Coronaviridae*), with their slimmed-down genomes, had perfected the art of entering their host's cells seamlessly, unlike large and bulky DNA viruses. mRNA vaccines were simply turning the tables and fighting fire with fire. "This is nature's doing; this is not really ours," he said. "We took a ride on something nature had figured out." It was a wildly imaginative solution, though neither Moderna nor any other vaccine maker had successfully brought an mRNA vaccine to market to prove the hypothesis.

Beyond its programmability, mRNA vaccines had one more potential advantage. Unlike other vaccine platforms, they weren't reliant on growing altered viruses (or parts of viruses) to inject into bodies, meaning they would be relatively cheap to produce—and fast: a few weeks was all that would be needed to create sufficient doses to vaccinate a large country's population, rather than the months or years it took to manufacture vaccines using traditional platforms, which relied on growing the viruses or viral proteins in large batches of organic material. The worst-case scenarios, though, were serious enough. Could the mRNA, given its fragility, simply break down once it was introduced to our immune systems, rendering itself unusable? Would mRNA collide with free-floating RNA strands present in our bodies, which research suggested might cause strokes? Would it produce a strong enough immune response? In 2015, though the potential for mRNA vaccines appeared limitless, so, too, did the doubts about whether the vaccines actually worked.

In 2017, Barney Graham and his team at the Vaccine Research Center were putting the finishing touches on a stabilized MERS spike protein. When it came time to test the new MERS vaccine, Graham turned to his institute's new partner, Moderna. The model was simple: Graham's team would provide the company with a blueprint for the stabilized MERS spike, and Moderna would produce an mRNA vaccine candidate that encoded for that spike. The next step was for someone to test whether it worked. Accurately testing a vaccine required access to systems (a range of human respiratory cells, mouse models, and other animal tissue) and an immersive knowledge of how coronaviruses attacked and entered cells. The tester also had to be able to run multiple flawless experiments simultaneously, and, preferably, be generally pleasant. Those demands added up to a very short list of potential collaborators.

"Ralph Baric is a brilliant, brilliant scientist," he said, "and really probably understands coronaviruses better than anybody on earth." And, Graham added for good measure, "he is just a real prince of a human being." Baric jumped at the chance to test a ready-made MERS vaccine and to see how well a new, synthetic technology performed against the ancient, organic tech burnished by his old adversary and muse. Soon after signing his contract in 2018, he started receiving shipments of vaccine doses sent from Graham's lab in Bethesda, Maryland.

Baric had years earlier created a MERS chimera by taking MHV, the mouse coronavirus, and synthetically grafting MERS spikes onto it. When Moderna's mRNA vaccine candidate arrived in his lab, he set to work infecting mice with his MERS chimera to set up a "lethal challenge" study, which involved injecting them with enough of the chimeric virus that they would die if they weren't given an effective cure. Baric had over the years seen a couple of promising SARS candidates and had even developed some himself; there were even fewer viable MERS vaccine candidates. (A review of the state of MERS vaccinology published in 2017 summed up the challenges neatly, which included "incomplete understanding of viral transmission . . . no optimal animal challenge models, lack of standardized immunological assays, and insufficient sustainable funding." Basically, everything.) The biggest problem was that coronavirus infections, MERS included, tended to be severer among people who were old, immunocompromised, or chronically unwell; even the most promising MERS vaccine candidates developed to date performed poorly among these groups, meaning that they would be ineffective at preventing death if the coronavirus became pandemic. If Moderna's vaccine performed as badly among those groups, it would be a much fancier version of the same old technologies that had failed to make it to market. But if it could protect the most vulnerable, then that would rank as a true discovery: a vaccine that didn't just train the immune system ahead of time to recognize and respond to an intruder, but that coached it to punch above its weight class, like an untrained amateur transformed into a pro boxer overnight. Baric hoped that this would be the case, and some of Moderna's other vaccine candidate data suggested as much. But this was the company's first time engaging with coronaviruses, which, though they appeared docile, were elusive and tenacious pathogens. Baric had reason to be skeptical. Still, as he readied his phalanx of mice for experimentation, he let himself hope that this was the breakthrough that would tip the scales back toward humanity after the *Coronaviridae*, in SARS and MERS, had announced its intention to create a new pathogenic age.

Baric's experiment, ostensibly set up to test a MERS mRNA vaccine, was actually testing two unknowns. The first was whether mRNA itself was a more efficient way of delivering vaccines into bodies than other means. The second was whether Graham and McLellan's pinioned spike protein performed better as a target compared to the wild-type spike. To get his answers, he set up a series of head-to-head vaccine experiments. The first of these compared the immune response in mice to Moderna's mRNA vaccine (which encoded instructions to produce the pinioned spike protein, but not the protein itself), to the immune response to a different vaccine that included the actual pinioned spike protein. This experiment was designed to test whether Moderna's delivery system really worked. The second experiment compared two mRNA MERS vaccines: one that used the pinioned spike and another that used the wild-type spike that wasn't held in place by prolines. This experiment was proof of concept for Graham and McLellan's theory that the shape of antigens mattered to how well vaccines performed.

Baric's team set to work, carefully selecting hundreds of mice and then randomly assigning them to one of the arms of the two experiments. Baric, as always, moved quickly, wasting no time to get the experiment up and running. But he still had to sit on his hands and wait: there was no way to rush how quickly the vaccines primed the immune system to counter pathogens. So, after an excruciatingly long six weeks—four weeks after the vaccine boosters had been given to the mice—Baric took blood samples from the mice and hunted for the presence of neutralizing antibodies. These microscopic munitions would signal that the mRNA vaccine had woken up the immune system to the menace and that it was responding with its full fury. If they weren't present, or were present only in low quantities, then all was lost. Moderna's vaccine, seeded with Graham's pinioned spike protein, would be yet another in a long line of failures.

What Baric saw made him laugh, so closely did it border on the magical. The MERS-infected mice injected with mRNA instructions encoded with the spike protein didn't get sick. What's more, they had neutralizing antibodies at far higher quantities compared to the mice injected directly with the pinioned spikes. Moderna's mRNA technology worked. And when Baric looked at what happened when the mRNA vaccine delivered a pinioned spike protein or a wild-type spike, he discovered that Graham and McLellan's hunch had been proven fabulously right as well. The "2P" spike produced more than twice as many antibodies as the wild type. It was a moment of unfettered success.

But the best part was yet to come. When Baric looked at the immune response among young and old mice, he couldn't detect a difference. Old mice were just as protected from MERS by the vaccine as younger ones. The combination of Moderna's mRNA technology, shepherded along by Zaks, and the artificial spike payload produced by McLellan and Graham at the NIH had done the impossible: it had created a vaccine that overcame the inevitable decline in immunity that caused so many older people to die from respiratory infections. Baric, who was rarely surprised by anything, admitted that he hadn't seen this coming. All the mice, regardless of age, displayed strong immunity. It was an unmitigated scientific breakthrough. Even so, Baric tempered his enthusiasm, no doubt recalling the classic virology joke: Humanity can cure every disease known to man—in mice.

Chapter 9

IT WAS WAR GAMING

RALPH BARIC'S BIOSAFETY LEVEL 3 CORONAVIRUS LABORATORY WAS located on the third floor of the Michael Hooker Research Center, at the far western edge of the University of North Carolina campus. Though out of the way, the building was conveniently located next to the MLK Jr. Boulevard entrance, allowing Baric to zip back and forth between his home and his lab at a moment's notice. It also happened to be next door to the UNC Eshelman School of Pharmacy, where, one day in 2018, two scientists were venting.

Tim Willson was a chemical biologist and former pharma scientist who had recently found his way back to academia. Nat Moorman was a microbiologist and close collaborator of Baric's. Willson had devoted years of his life to the discovery of kinases, enzymes that regulate cell functions and that viruses need in order to attack human hosts. He had been recruited to UNC in 2015 to lead the laboratory for the Structural Genomics Consortium, an international network of scientists who mapped how proteins produced by the body were involved in immune response. Moorman, who had been at UNC for almost twenty years, studied mostly cytomegaloviruses, a cluster of viruses closely related to herpes. Though they worked in different fields, both were frustrated at the slow and myopic way science seemed to be moving in the twenty-first century. It wasn't so long ago that they could remember working on creative, open-ended collaborations with scientists across disciplines, each new voice feeding into the energy of discovery. Both itched for something more. As they chatted, Moorman mused to Willson about the possibility of targeting kinases as a way to stop cytomegaloviruses from replicating. Willson thought about it for a moment, then responded that there was no reason to stop there. So many different viruses needed human kinases to replicate that if you were able to remove the tiny bundles of molecules, you might be able to stop viral replication across the board, no matter which viral infection you were dealing with. It was an exciting prospect.

The interdisciplinary collaborations that Moorman and Willson yearned for were the centerpiece of Ralph Baric's career. There was a reason that Barney Graham called Baric "a prince of a human being": the coronavirologist had made a point of working with his friends. Baric and Mark Denison, the soft-spoken pediatric virologist, had spent thirty years hunting the virome together, despite working at separate institutions in different cities. (Denison worked at Vanderbilt University in Nashville, Tennessee.) Their collaboration was so devoid of ego, in fact, that Baric couldn't remember which of them had first originated many of their ideas.

Nearer to home, at the University of North Carolina, Baric drew close to an ever-growing cadre of virologists, all of whom were pursuing their own pet viruses, but who invariably were pulled into Baric's orbit on account of his dogged pursuit of virological truths and his scientific excellence. (In some cases, devotion to Baric rose to the level of self-sacrifice. One of his UNC collaborators, the virologist Mark Heise, ceded his entire laboratory, which studied mosquito-borne viruses, to Baric so that he could expand his coronavirus research program. "To be frank," Heise explained, "it's really because Ralph and I are friends. Ralph and I enjoy collaborating with each other, and we have similar thinking about how we treat people.") Moorman was among those who evidently loved Baric as both a person and a scientist; the two of them had often gone down virological rabbit holes together for the simple reason that pursuing knowledge was joyful.

Moorman's conversation with Willson had opened up a wide vista stretching far into the future. If the two of them were right and blocking kinases could stop viral replication, then they might have just landed on a strategy to disrupt whole families of viruses. It was exciting, and though the details of how they would do it were still fuzzy, Moorman knew one thing for sure: Baric would be intrigued. So, he tracked his friend down and excitedly told him all about the plan to shift the virological battlefield from individual viruses to entire viral families. What's more, Moorman explained that he and Willson had even settled on their first targets: alphaviruses and flaviviruses, families that had been largely ignored by the private market but whose members (Chikungunya, Dengue, and Zika) were responsible for tens of thousands of deaths in poor countries every year.

Baric loved the idea. One of the chief reasons was, of course, how closely it mirrored the work that he and Denison had been doing for years. But Baric was also glad to have someone to share the gnawing anxiety he had been unable to shake since MERS emerged. It was such a shame, the two virologists agreed, that every time an epidemic hit, the world was in reactive mode. Then, in the space of a year, it would all go away again as funders and scientists got distracted by the next big thing. Baric was all in on joining Moorman and Willson's nascent project. He had only one suggestion: include the *Coronaviridae* alongside alphaviruses and flaviviruses. Moorman agreed; in fact, it was exactly what he wanted to hear. He was well aware of Baric's years-long search for a pan-coronavirus killer, which had culminated in the discovery that remdesivir and EIDD-2801 could elude the coronavirus proofreader.

The plan felt big and important, but beyond that feeling, none of the three really knew what to do next. Baric, Moorman, and Willson were all basic scientists, fluid in the lab but fuzzier on strategy. So, while the research made sense—the number of novel viruses with pandemic potential was only speeding up as the twenty-first century unfolded—they weren't sure how to translate it into the real world. They needed someone else—preferably, a scientist who understood the basics of molecular biology just like them, but who had found a way to get that knowledge out into the real world, where it could do some good.

After some thought, Willson piped up. There was someone, actually, who hit all the marks. His name was Aled Edwards, and he was Willson's boss at the Structural Genomics Consortium. Edwards, Willson was sure, would know exactly what to do. After all, he had built his career by doggedly pursuing the big ideas he believed in; the only hitch was that he also worked feverishly to destroy the ones he opposed.

ALED EDWARDS HAD FOUNDED THE STRUCTURAL GENOMICS CONSOR-tium in 2003 to great fanfare. A tall, thickly built man with brown hair and an imposing demeanor, Edwards was a motormouth with a penchant for spiraling into emotional tirades; he had no use, it seemed, for self-censorship. His face was a constantly shifting range of emotions, his thick eyebrows and deeply creased smile lines broadcasting profound feelings of accomplishment, confidence, joy, grief, and fury, sometimes in rapid succession. Edwards had built the Structural Genomics Consortium with a clear scientific directive in mind: to create three-dimensional maps of the scores of proteins the human body produces. But the consortium was also Edwards's path to achieving his dream of tearing down the capitalist structures that excluded most of the world's population from accessing expensive medicines, and replacing it with a system of science for the public good. Curiously, that dream was financed by the bottom line of pharmaceutical giants, for whom mapping proteins was critical to their ability to make drugs and the profits that came along with them.

In 2003, while pharma profits were as high as ever—between 1990 and 2002, U.S. prescription drug sales more than tripled from about $45 billion to almost $200 billion—companies were seeing their novel drug pipelines stall out. It boiled down to a market paradox: most drug development started by identifying target proteins in the body, which the drug could act on in order to prevent illness. But because the body produced so many proteins, pharma companies couldn't justify funding the interminable work of mapping them. The human body produces, by some estimates, as many as four hundred thousand proteins, the vast majority of which we know nothing about, while mapping just a single protein can cost upward of one million dollars. Wading through the universe of proteins is like trying to identify asteroids in the Kuiper Belt: they all look the same, there's no end in sight, and there's no way to know which of them are worth anything. At some point in the late 1990s, fed up with the paradox, some pharma companies started looking for a path out.

Soon, scientists at GlaxoSmithKline decided that it would be a whole lot easier if pharma companies pooled money to pay a non-profit third party to do the protein mapping and then shared the maps among themselves. Others agreed and the group started looking for an organization that could organize the venture, landing on the Wellcome Trust, a United Kingdom–based charity that was one of the main funders of scientific research around the world. Wellcome, happy to receive millions in pharma dollars to fund a project with the potential to accelerate the manufacture of drugs to save lives, agreed. The next step was to find a leader. The person needed to be a heavy-duty structural biologist with a track record for outsize success in both the public and private sectors. Aled Edwards fit the bill—assuming he didn't inadvertently burn the whole thing down.

By 2003, Edwards had a reputation for two things: scientific excellence and bomb throwing. He had trained as a postdoc at Stanford under the Nobel Prize–winning chemist Roger Kornberg, who was interested in how DNA went about replicating itself, a process known as transcription. Edwards was exquisitely suited to the work, which involved probing the lattice-like three-dimensional crystalline structure of proteins involved in transcription, and he was the first author on one of the key papers the Nobel jury cited in awarding Kornberg the prize. After his stint at Stanford, Edwards returned to Canada and ended up on the faculty at the University of

Toronto, where he landed as a young superstar ready to show off his scientific chops. He was intent on taking his immersive understanding of structural biology and using it to produce cures for neglected diseases. But just as his career was on the upswing, he came face-to-face with a problem bigger than his scientific dreams: the decades of massive government spending on STEM (science, technology, engineering, and mathematics) were abruptly drawing to a close.

After peaking in the late 1980s, government R&D funding across North America shrank precipitously in the early 1990s, a trend that would extend until the dawn of the twenty-first century. Right when Edwards was trying to spin his impressive scientific pedigree into research gold, the bottom suddenly dropped out of the public market. From a high of 1.2 percent of GDP in 1987, U.S. government expenditures in science dropped by over a third, to 0.8 percent in 2000, forcing thousands of scientists who relied on public grants to scramble for dollars elsewhere. Edwards, like so many, hit a wall. His scientific discoveries were making it into *Nature*, the world's most prestigious scientific journal, but his experiments were being run on thirty-thousand-dollar government grants, which barely covered keeping the lights on at his lab. There had to be a better way.

While public research funding was on the decline, the biotech market was in full swing. By the late 1990s, the industry was already worth well over one hundred billion dollars, and investors were pouring in. It seemed stupid to Edwards to ignore biotech's obvious competitive advantage. So he jumped into the fray and launched his own company to develop antibiotics for rare diseases. Then he launched another company, and another, all of which paid off handsomely. He was a natural.

It wasn't long, though, before he started to feel twinges of concern. While the financial success of his companies provided him the space to flex his scientific muscles, it also gave him a rough education in the darker side of scientific capitalism. One of the first major hires he made after his initial foray into the market was the CEO who replaced him; the man was a lawyer whom Edwards credited with teaching him "aggressive patent law." This meant filing patents on every single discovery, no matter how mundane. The point, Edwards was taught, wasn't to protect yourself from a competitor who might infringe on your hard-earned scientific labor; the patent existed as a negotiating tactic that could be used in future deals. Edwards was schooled in the dark art of using patents as weapons in business, which time and time again the lawyer CEO used to outflank competitors and protect his investment. At first, Edwards found this riveting; it quickly became sickening. What was the point of a patent if it was going to slow progress instead of speed it up? "This is just fucking stupid," he recalled thinking at the time.

In 2002, after his journey through the darkness of for-profit science, he returned to academia a changed man. He was no longer a believer in the scientific paradigm of profits and patents: he had become an iconoclast, obsessed with tearing the whole thing down and replacing it with a model of open science—no patents, no marketplace, just drug discovery that met the needs of all the earth's peoples, no matter where they lived or how much money they made. All he needed was a platform to do it.

When Wellcome approached Edwards in 2003 with an offer to lead a Big Pharma protein-mapping enterprise, he leapt at the chance. Though it was funded by massive private interests, Edwards figured that the initiative, which he dubbed the Structural Genomics Consortium, was the ideal vehicle to launch his open-science revolution. He quickly laid down some ground rules. The first was that nothing would be patented: all the protein structures the consortium mapped on behalf of pharma would be made publicly available in a government-funded database where anybody could access them. The second was that the pharma companies funding the enterprise would get only a few special perks (early access to some protein maps and a wish list of specific proteins they could submit) and nothing more. The first company to formally jump on board was GlaxoSmithKline, and nine others joined soon after, among them industry heavyweights like Bayer, Pfizer, Janssen, Merck, and Novartis. By 2015, Edwards had amassed close to one hundred million dollars in funding, including sizable sums from the Bill and Melinda Gates Foundation. And he had labs across three different continents: in Toronto, Oxford, Stockholm, Frankfurt, the Brazilian town of Campinas, and (last to join) Chapel Hill, North Carolina, where he employed hundreds of scientists, Tim Willson among them. The consortium was, by all measures (financial, scientific, the public good) a smashing success. For Edwards, it wasn't enough.

From the start, it was clear that Edwards relished the idea of being a visionary. His stated mission was to transform the financing of modern scientific discovery so that it didn't disenfranchise millions of the world's poorest. He was perhaps the world's only working purist on the subject of open science and the best funded of them all. It is a peculiar purist, though, who works to advance his anticapitalist agenda by using millions of dollars from Big Pharma. By 2018, the price tag for a pharmaceutical company to join Edwards's Structural Genomics Consortium was eight million dollars, a significant sum even for companies that measure their revenues in the billions—especially when they were writing their checks to a man who made no attempt to hide his disdain for the market and its players.

Edwards had found a way to successfully walk a long and shaky tightrope between open-science principles and the interests of pharmaceutical giants. With the consortium thriving, he figured it was time to go deeper. If the open-science model worked, why stop at mapping proteins? The private sector wasn't going to make new drugs for rare diseases and those that killed poor people in lower-income countries. And if they weren't going to, Edwards figured, the consortium would have to get into the drug development game and do it for them, one way or another.

WILLSON AND MOORMAN CAME TO EDWARDS WITH THEIR IDEA, WHICH was to develop drugs that could shut down entire viral families by targeting the proteins produced by the human body that the viruses needed to replicate. Always the visionary, Edwards immediately grasped the scientific importance of Willson and Moorman's idea, but he told them to think bigger. How could Moorman, Willson, and Baric take the idea and turn it into a platform to end viruses forever? And how were they going to take their idea and transform it into something real?

The first order of business was to come up with a name. Edwards, keenly aware of the need for good branding, helped them land on READDI, for "Rapidly Emerging Antiviral Drug Development Initiative." While READDI's ostensible mission was to bolster humanity against future pathogens, it also fulfilled Edwards's ardent desire to transform the market-driven paradigm of discovery. "READDI emerged from watching how the world responded to malaria, tuberculosis, and all those other neglected diseases, all of which suffer from no market incentive to invent a medicine. We have all these diseases where clearly it doesn't work to think of capitalism."

Edwards didn't begrudge the pharmaceutical industry for its lack of incentive, though. "I absolutely do not shit on pharma. Why should it be Merck's job to save the world? It makes no sense. They're just doing what we set corporations up to do." It was the fetid mix of hubris on the part of pharma companies and debased humility on the part of everyone else that had led society to accept that only large, private companies with a fiduciary duty to maximize profits could make earth-shattering discoveries. "We need to invent a different way," Edwards said flatly. "Let's take advantage of what the private sector is good at, which is the rigorous processes, the assembly lines, all the stuff that is hard to do in a public sector. Then let's take the innovation part and bring it back" to the public sector.

Still, with Edwards in Toronto and occupied with running his global open-science empire, READDI remained more artful semantics than reality. It was a nice program among many at UNC, but like many academic pursuits, it was hard to know exactly what it was accomplishing. By the summer of 2019, Willson, Moorman, and Baric had garnered some modest financial support from the Eshelman Institute for Innovation, a division of the UNC Eshelman School of Pharmacy. This money had allowed them to devote a few hours a week to moving the project forward. Outside their labs, though, READDI was nothing more than a concept. No one had yet mustered the energy even to create a website, let alone a bona fide organization to turn their findings into a real-world solution. Why would they? READDI was largely theoretical, the kind of science that scientists love but that most people (and funders) dismissed as largely irrelevant to the here and now. Until the day that the future threat that READDI had been preparing for landed in the here and now.

September 2019, Bethesda, Maryland

Anthony Fauci, Tal Zaks, Barney Graham, John Mascola (Graham's director), and Moderna CEO Stéphane Bancel were huddled around a conference table, talking about viruses and vaccines. It was the annual meeting between Moderna and the National Institute of Allergy and Infectious Diseases (known as NIAID), which Fauci led, to discuss how their collaboration was progressing. It had been a good few years. Between Moderna's continual refinement of its mRNA vaccine platform and NIAID's in-house expertise in identifying priority pathogens and running clinical trials, the two sides had helped close the gap on vaccines for some of the most elusive and deadly viruses the virome had foisted on humanity.

Zaks had been asked to present some of Moderna's recent projects, chief among them a vaccine for cytomegalovirus, a vexing pathogen that used a pentameric antigen (akin to a five-sided spike protein) to attack and enter cells. Traditional vaccines that tried to introduce the entire pentameric antigen into cells had failed because the antigen kept falling apart. Moderna's solution to the five-sided problem was sublime. Under Zaks's direction, the company's vaccine team created five separate strands of mRNA that encoded for each of the five sections of the cytomegalovirus antigen. Once delivered inside a cell, each mRNA strand instructed the cell to make a different protein that made up the pentameric antigen. The five proteins then folded together inside cells as if by magic, triggering an immune response. It was the height of scientific elegance. Between the vaccines that Moderna had already developed and the apparently limitless potential of mRNA, the NIAID team, including Fauci, were impressed. The conversation soon turned to what to do next.

Despite not having yet brought a vaccine to market, Moderna had spent the past four years establishing proof of concept for their mRNA vaccine program. The assembled scientists all agreed that their collaboration, in which Moderna made vaccine candidates and NIAID ran trials to see if they worked, had turned out great. But when a pandemic-ready pathogen emerged, the question wouldn't be only whether a vaccine could be made, but also how quickly. So, the NIAID team proposed a race. First, the institute would pick a virus. Then, on a given day, the full genome sequence of that virus would be sent to Moderna. At that point, Zaks said, "the artificial clock would start, and they would see how fast we can move and produce it, and how fast we can get a Phase One clinical trial done." Zaks and Bancel, who had made their careers delivering products on deadline in the pharma market, loved the idea.

Over the next three months, Zaks, Graham, and the others went back and forth on the specifics: which virus to choose, how to structure the challenge, and how much money the whole enterprise would cost. By early November 2019, they had settled on the Nipah virus as the target, and all sides were readying themselves to start the artificial clock ticking. "It was war gaming," said Zaks, with delight, a chance to test humanity's battle readiness before the live fire began.

They would never get the chance to run the drill.

Part III

Chapter 10

YOU EXPECT HEROES TO DIE

O N December 31, 2019, the Chinese government reported a cluster of twenty-seven unusual pneumonia cases in the industrial city of Wuhan. Four days earlier, a partial sequence from one of the patients had suggested that the cause of the illness was a new coronavirus. China's National Health Commission urged restraint. In terse statements, Wuhan city health officials toed the party line, describing the cases as under control and insisting that no sign of human-to-human transmission had been recorded. Despite the rumors around the origins of the virus, officials insisted, "We've investigated, and it has no relationship to SARS."

Ian Lipkin had seen this story before. As soon as news leaked of the Wuhan pneumonia cluster, an illness characterized by strange white spots in the lungs, Lipkin's ears pricked up. When early tests suggested that the pneumonia was likely caused by a coronavirus, he was rapt. He also knew that time was of the essence. The sooner the Chinese government instituted a basic outbreak response (closing outbreak sites, restricting population movement, scaling up testing, and initiating contact tracing), the higher the chances were that the new virus could be contained. Every day that those responses were delayed meant one more day the virus had to spread. Lipkin was unswayed by official statements. As long as new cases were being reported, the outbreak was evidently not under control, no matter how many times public health officials downplayed its severity.

For all his anxiety, Lipkin wasn't quite sure what to believe by the end of December 2019. "It wasn't clear that there was a lot of human-to-human transmission," he admitted. That wasn't as surprising as it may seem. During the first SARS epidemic, the period when afflicted people were infectious and when they developed symptoms happened at exactly the same time, and lasted only thirty-six hours. This made it relatively easy to do contact tracing and find out who was transmitting the virus to others. In the first days of the Wuhan pneumonia outbreak, the only clue was that, early on, many of the infected had visited the Huanan Seafood Market, which specialized in live-animal sales. By early January 2020, though, patients who hadn't visited the market started presenting with fever at Wuhan-area hospitals. And still, calls for the government to respond forcefully were largely going unheeded. Lipkin had an idea of what was happening. "There were people in China who I believe honestly thought that this virus was under control," he said. "But I don't think they were evil in that way; I think they were in error, and it was a grievous error." Away from the anodyne statements repeated under the glaring spotlight, Lipkin had been privately told by his Chinese contacts that the government was sourcing PPE and emphasizing physical distancing.

For all their novelty, epidemics are predictable forces. The epidemic triangle (the shifting relationship among a pathogen, its host, and the environment within which the relationship plays out) can always lead you to the origins of an outbreak if you search hard enough. Questions about origins were, Lipkin agreed, important: the best way to stop an emergent pathogen was by tracing back the filaments of its trajectory to its ultimate source and then sealing the fissure from which it first emerged. But there were more pressing concerns in the short term: the world needed to rapidly set in motion the scientific response (testing, therapeutics, and vaccines) in case a pandemic-ready pathogen had indeed emerged in Wuhan. "Some people think it's very important to figure out where it came from," he said, "but I'm actually less concerned about that than I am about what we are going to do about it." Regardless of the order of priorities, tracing the origin and kick-starting the scientific response required the same thing: for Chinese scientists to sequence the new coronavirus's genome and share it with the world.

Thanks to his epidemiologic prowess and code of discretion, Lipkin was among the best-connected scientists in the world, with a Rolodex that included senior members of the U.S. State Department and the Chinese Communist Party. In a 2016 trip to China, he was given a medal by Chinese president Xi Jinping in honor of his work helping the country respond to SARS. During that trip, Lipkin also reconnected with many of the scientists and party officials he had collaborated with on his first fateful trip to Beijing in 2003 during the SARS epidemic. By January 2020, many of those contacts were now senior figures in government, including Chinese premier Li Keqiang, the second in command to President Xi. In January 2020, with the Wuhan cluster spreading, Lipkin pleaded with Li to act more aggressively to contain the new coronavirus, and he urged the government to allow Chinese scientists to publicly release its genomic sequence as soon as possible.

By January 5, that genome mapping had been completed. Unlike the week that it took Brunham and Marra to map the SARS virus, it had taken Chinese scientists a matter of hours. But while the science had advanced dramatically since 2003, the politics hadn't. As soon as the sequence was mapped, a government memorandum ordered virologists not to share it without official authorization. This led to more delays and days lost during which the global scientific community was kept in the dark about the nature of the threat it faced.

On January 10, 2020, when fewer than fifty official cases had been identified, the genomic sequence of the novel coronavirus that would ultimately be known as SARS-CoV-2 was finally released to the world. It was a pointillist masterpiece, a portrait thousands of molecules long, capturing the curious new threat from Wuhan. Fixed in place and held together by atomic energies, the molecules arranged themselves, following one another, along the single helix of RNA. The script was made up of four molecules (adenine, guanine, cytosine, uracil) repeated, like an evolving mantra, over and over again, the string reaching a thousand nucleotides and then a thousand more and, finally, almost thirty thousand in all. The sequence, a secret code written in simple script, constituted a guidebook. It told the story of the new enemy of humanity, an elegantly minimal entity that existed at the break point between living organism and automaton.

On January 11, 2020, when he saw the sequence published in the NIH's public access database, Lipkin was relieved. The release of the sequence meant that the forces of public health in China were winning the day against those in government more concerned with saving face. Still, the sequence was only the beginning of the story. Lipkin was a shoe-leather epidemiologist, convinced that if you wanted to understand an epidemic, nothing was as good as being there. Relying on official communiqués about progress against an epidemic was foolhardy. You had to be in the room. Lipkin had also gravitated toward an elder statesman role that he had earned over his years as a globe-trotting epidemiologist. When epidemics arise, he said simply, "I advise."

In mid-January 2020, Lipkin flew to Guangzhou, the southern port city where the SARS epidemic began in 2002. Traveling to Wuhan, at the time, was a nonstarter. Information blackouts, along with cordons of police officers and public health officials, had made the city largely impenetrable to outsiders. No matter: Lipkin was in China to play the role of honest broker with those who controlled the upper levers of government. When he landed in Guangzhou, he met with the One Health research team with which he had long collaborated, which was made up of scientists who investigated epidemics by tracking the increasingly perverse relationship between human, animal, and environmental health. The emergence of yet another novel coronavirus—at a live-animal market, no less—had renewed their resolve to cut the head off the snake. In his meetings with Chinese government officials, Lipkin ran with that message: the markets had to close. There was pushback, of course, but he had expected that. China's exotic wildlife market was worth upward of $76 billion annually, with the exotic food sector valued at $20 billion. Money and politics, he knew from his dealings with the Saudi government during MERS, would always upend the best-laid epidemic-containment plans. But responding to a zoonotic epidemic while keeping the wild markets open was like bailing water out of a sinking ship without first plugging the leak. Lipkin made his case to his many contacts in China's Ministry of Science and Technology, the Ministry of Health, and the central government (including Premier Li Keqiang). Sensing the gravity of the situation, they agreed.

By mid-January, Wuhan's wildlife markets had closed, and on February 24, 2020, the Standing Committee of the National People's Congress had enacted a total ban on the consumption of wild animals in China. Lipkin also helped with less dramatic but equally important issues, like selecting the right diagnostic tests and serologic assays to screen for the novel pathogen. At a time when the world was flying blind in the face of a new viral threat, he also emphasized using convalescent plasma therapy (blood from people who had been infected with the coronavirus and had survived) as a promising early-phase treatment. "Those things I did may not have made a difference immediately, although I think they probably did," he said. He had no doubt about one thing, though. "It will make a difference in the long term."

In the short term, the glimpses of the Wuhan epidemic appeared increasingly otherworldly: Public health officials in hazmat suits pulled gurneys through the streets. An elderly man, dressed neatly in a black suit and wearing a mask, lay on his back on the pavement, dead, a white plastic shopping bag clutched in his hand. An ophthalmologist, Dr. Li Wenliang, pleaded into his phone camera that the crisis was much more dire than anyone had been led to believe. (The same Wenliang succumbed to the virus at the age of thirty-four after being forced to sign a confession for "publishing fictitious discourse," the memory of him furiously scrubbed by state censors.)

And then came the news that the industrial city's eleven million residents would be subject to a full lockdown, which seemed, in January 2020, an unfathomably draconian response to what was surely a redux of the SARS blip.

January 13, 2020, Cambridge, Massachusetts

Tal Zaks, Moderna's chief medical officer, had been eagerly awaiting the start of the vaccine war games. But as Moderna and the National Institute of Allergy and Infectious Diseases negotiated the details, a real live enemy materialized. "We were planning to do this thing in 2020," Zaks said, "and then in December, you start hearing about people coughing in Wuhan." In early January, when experts like Anthony Fauci were telling Americans that there was at present no reason to be concerned, Zaks had been hearing a different message: "This is not a drill, this is actually live fire, and it's time to muster up."

Though the pandemic-ready pathogen arrived earlier than they would have liked, it had the effect of focusing, rather than undoing, the meticulous plan the team (including Zaks, Fauci, Graham, and others) had developed. With the broad strokes of their war games agreement already in place, both sides were primed for launch. The drill would have seen NIAID start the ticking clock by sending Moderna a viral sequence, which the company would then program into a vaccine and start production on doses. NIAID would then initiate a Phase 1 clinical trial as fast as possible, to test whether the vaccine worked. With the hazy reports of the novel coronavirus cohering into a picture of a formidable pathogen capable of extreme illness and efficient, asymptomatic transmission, only one thing had changed: NIAID was no longer going to choose the virus nor set the clock. Instead, it would start on Friday, January 10, 2020, when the sequence of the novel coronavirus was uploaded and shared with the world.

Graham wasted no time. Early the following morning, he headed to his lab at the Vaccine Research Center in Bethesda, Maryland,

and spent the rest of the weekend holed up there, analyzing the sequence. By then, only forty-one cases of the novel coronavirus had been confirmed, all in China, and only one infected person had died. This low prevalence was no deterrent; Graham had spent his entire career tracking emerging epidemics, and they had all, by definition, started small. By Sunday evening, he and his team had isolated the spike protein sequence and had stabilized it using their 2P adjustment, which held it in the shape it had when it first entered the body, which boosted the immune response that a vaccine produced. By Monday morning, while the rest of the world was still conducting business as usual, Graham sent the stabilized sequence to Zaks and the Moderna group. That same day—January 13, just three days after the public release of the genome—Moderna launched production of mRNA-1273, its COVID-19 vaccine.

It was, despite the unfolding terror of the pandemic, the best-case scenario. Moderna and Graham's team had already produced a MERS vaccine candidate that Baric had shown effectively protected both young and old mice from illness (though the slow spread of MERS had left them with too few numbers for a clinical trial among humans). The vaccine that Zaks's team designed for the novel coronavirus was essentially identical, except that the mRNA strand encoded the new virus's spike protein instead of a MERS spike protein. Though there was no way they could get it out to the public before human trials were run to test whether it was well tolerated and effective, Moderna nevertheless had a working vaccine as of early January. If the novel coronavirus remained a small and contained outbreak, as Chinese officials suggested it would, then the vaccine would all be for naught; Graham and Zaks would run into the same numbers problem that had crashed Bob Brunham's dreams of getting a SARS vaccine to market.

Compared to the five-pronged mRNA vaccine Zaks and his team had put together for cytomegalovirus, "this one was easy," he said, referring to the novel coronavirus vaccine. "I mean, my researchers hate me when I say it, but at the end of the day, it was kind of easy." What made it so, beyond Moderna's capacity to rapidly program mRNA strands to deliver essentially any instructions to human DNA, was the public release of the genome and Baric's research demonstrating the importance of the spike to human infection. Zaks had particular reverence for the four decades of scientific labor Baric had undertaken. "It's not by chance," Zaks explained, that Moderna and Graham's group at the NIH knew that they should include the coronavirus's spike protein. "The world understood that the spike protein was the most relevant one." There was no need to hunt around for the right antigen; Baric had been describing it for years.

But even though Zaks was able to oversee production of mRNA-1273 a mere four days after the novel coronavirus genome was released, a vast gulf stood in the way of getting it to market—namely, evidence that it was tolerated and effective—along with less scientific challenges.

Zaks minimized the role of market forces in motivating his team. But when he spoke about Moderna's credibility deficit—here was a young company with a radical, unproven vaccine technology—it was a subtle but undeniable nod to his competitors. The main one was Pfizer, the only other major pharma company that had developed an mRNA vaccine candidate. Pfizer had partnered with BioNTech, a German biotech company that had brought in Drew Weissman and Katalin Karikó (the virologists who originally inspired Derrick Rossi, the founder of Moderna, to experiment with mRNA as a way to deliver protein-building instructions directly to DNA) to lead the work. Moderna's only hope of competing with the brand-name recognition of Pfizer, Zaks well knew, was to be among the first to market. Moderna, with the imprimatur of the NIH running a clinical trial for its vaccine candidate, could afford to be a new entrant at the head of the pack. But if the seasoned players outpaced it, the company would never be able to regain its footing, no matter how effective mRNA-1273 turned out to be.

Zaks organized to have Moderna's initial batch of vaccines sent to Graham at the Vaccine Research Center. Graham, just as they had planned in the war game, then had the responsibility of running the various trials that would prove to the world that the technology actually worked. There was no time to waste. To get vaccines to market as quickly as possible, the team had decided to run animal trials (to test whether the vaccine prevented infection and disease progression) with a Phase 1 human trial (which tested safety among a small number of participants) at the same time rather than sequentially. If everything went off without a hitch, Graham reasoned, then by July 1, 2020, the NIH could launch a combined Phase 2/3 clinical trial among tens of thousands of people to test the vaccine's efficacy and its adverse effects. In the very best-case scenario, that could get vaccines into arms by the end of the calendar year. It was an aggressive time line, but Graham felt it could be done. He also knew, though, that the July 1 launch date was dead in the water if he didn't have high quality and highly persuasive animal data in hand. Luckily, he had a friend who could help.

With the pandemic doomsday clock potentially striking midnight, Baric was informed that the future of humanity was suddenly relying on his expertise with mice. Graham needed mouse model data showing that the mRNA-1273 vaccine was tolerated and effective, and he needed it in under six months. Graham was steadfastly against rushing the process and risking repeating the nightmare scenario of the RSV trials of the 1960s, when the vaccines ended up killing infants rather than saving them. The lethal challenge studies that Baric ran were critical to making sure this didn't happen. And that meant Baric would have to create genetically modified strains of SARS-CoV-2 adapted to mice and dose them with enough synthetic pathogen that, if the NIAID/Moderna vaccine failed, the strains would die. Baric, resolutely committed to wiping out the family that had become his life's obsession, fetched his mice.

January 27, 2020, University of North Carolina, Chapel Hill

With a raft of urgent experiments under way, all seeking to find and exploit weaknesses in the novel coronavirus, Baric walked slowly out the doors of his lab like a stranger. His lab coat was open, revealing a blue golf shirt that fit tightly over his sturdy frame. The camera followed him as he advanced down the hall, his arms hanging awkwardly at his sides, and then caught him as he furtively looked directly at its constant gaze.

This was not the image of a man at peace: it was Baric coming to terms with the reality that his worst nightmare was coming true. "It is of concern," he said in his slow, gravelly baritone, his gray-blue eyes staring intently at the journalist from WRAL, the local Fox affiliate. "It doesn't mean you should be panicked or"—he visibly searched for the appropriate words—"overly alarmed. But you should be aware." It was clear that these kinds of soothing television appearances hadn't ever been a part of his scientific repertoire. In private, he exuded an easy confidence at stark odds with the controlled, affectless version of himself that he shared with reporters. And even as he warned viewers not to overreact, he knew that Moderna had begun production of its vaccine, while he himself had already, at Graham's request, initiated the lethal challenge study in mice.

As Baric soberly looked into the camera, trying and failing to convey a sense of calm, it was evident that the conflict between sickness and exhilaration was as alive as ever. This time, though, the early warnings from Wuhan suggested that this virus was something far worse than what had come before. Both SARS and MERS, while deadly, had also provided epidemiologists with a lifeline: more often than not, the transmission chains linking infections were clear. Patients, gravely ill, came to hospitals, and the nurses and physicians tending to them became infected, too. But so far, the reports from Wuhan suggested that this wasn't the case. The transmission chains seemed broken. People who had no contact with visibly ill patients were nevertheless ending up in hospitals with fevers, their lungs filling up with strange white spots and then their bodies losing their capacity to take in oxygen. It was a paroxysm of illness and, as January wore on, death—and still, contact tracing, which should have zeroed in and capped outbreaks shortly after they emerged, kept leading public health officials down blind alleys and dead ends.

As he told viewers to remain calm, Baric was aware that all was not well. The day before, he had watched in horror as China's health minister made the startling admission that the novel coronavirus could spread asymptomatically. This was a radical departure from both SARS and MERS. The novel virus was being transmitted more like the flu than like any other coronavirus Baric had encountered. "Once it emerges," he explained privately, "you have all kinds of people transmitting before they get sick, and so contact tracing's going to fail, quarantine's going to fail, unless you take on a draconian measure to quarantine sixty-five million people and use the military to enforce it." Even then, Baric knew, it wouldn't be enough to stave off the inevitable. In November 2002, the first SARS cases were reported in the Chinese city of Guangzhou; by March 2003, the disease had spread across the world. In March 2009, a new strain of swine flu was reported in the Mexican state of Veracruz; by June of that same year, the World Health Organization had declared a worldwide pandemic. In January 2020, Baric saw the writing on the wall. "The U.S.," he said, "had three months." After that, the virus would be everywhere.

Baric was once again contending with the psychic pain of knowing that people were dying from the organism he found more fascinating than anything in the world. It was a morbid fact of his profession, which tempered the exuberance of discovery with the reality of death. The more complicated the machinery of the problem with which he grappled, the greater its capacity to elude detection, to sicken, and to kill. He had spent four decades trying to work through this moral dilemma and hadn't quite gotten there. But this time, at least, humanity was not starting from zero. There were vaccines in production, something that hadn't ever happened with SARS and MERS. And there was something else: remdesivir and EIDD-2801, two broad-based antivirals that caused error catastrophe in every single coronavirus strain he and Denison had tested it against and that should, if the theory held, destroy this new pathogen as well.

"WE ARE UNDERPREPARED FOR THIS," THOUGHT ALED EDWARDS, AS the COVID-19 pandemic began its cruel assault on humanity. He wasn't thinking about the need for a specific vaccine or antiviral, though; that kind of small-mindedness was what had gotten us here in the first place. And it wasn't just that governments were revealing how poorly equipped they were to respond to crisis. Edwards had his eye fixed a magnitude higher. The greatest failure was humanity's collective faith in an archaic system of for-profit science to help save human lives. That system had become so deeply perverted by the ruthless complexities of capitalism that it had undergone its own version of error catastrophe: a million tiny pinprick pressures had forced publicly funded efforts to collapse, to the point that government was completely unable to create viable vaccines or therapeutic drugs without the private sector doing most of the heavy lifting. Even the NIH's partnership with Moderna was yet another example of government ceding the work of innovation to for-profit entities. What was needed—what Edwards had been railing about for years, according to anyone who had been cornered by him—was to wipe the slate clean. It was time to go back to the first principles of scientific discovery to save the most human lives possible. The effectiveness of a vaccine or antiviral in blunting the threat from the novel coronavirus would all be for naught if those medicines couldn't reach every single person in the world who needed them. You can't end a pandemic by protecting only the richest people in the world.

The expanding epidemic and the fear that it provoked was, Edwards believed, a clear opportunity to transform his decades-old dreams for a new kind of science into a reality. With the creeping realization that the new virus had pandemic potential, the global scientific discourse had abruptly shifted from words like *patents* and *intellectual property* to a simple, universal mantra: Sharing data is good. Here was an axial moment to bend the course of scientific discovery back toward the greater good. The only obstacles were the crushing weight of global capitalism and Edwards's inability to shut the hell up.

In February 2020, the Structural Genomics Consortium held

one of its regularly scheduled board meetings. The board was made up of ten different pharmaceutical executives representing the companies (including Pfizer, Johnson and Johnson, and Merck) that had paid to have the consortium's scientists map proteins that the companies then used as targets for the drugs they developed and sold. Edwards and Tim Willson both attended. So did John Bamforth, a former pharma executive and the new director of the University of North Carolina's Eshelman Institute for Innovation, which had given some modest seed funding to READDI.

The mission statement of Bamforth's institute was full of the buzzwords that were, for Edwards, signals that a university was doing science for the wrong reasons. The institute, its website stated, was focused on "impact" (Edwards, mockingly: "Look to our revenues for the impact we're having"), "economic development" ("Invest in university research, we'll generate the economic value of the future"), and, of course, "innovation" ("'The most innovative ...' What that basically means is counting patents"). Nevertheless, when Bamforth arrived at the University of North Carolina, Edwards invited him to become a board member of his open-science consortium.

There was, on paper, no worse match. Bamforth, had trained as a pharmacist and then had risen through the ranks to become the chief marketing officer at the pharma giant Eli Lilly. That practically made him the antithesis of open science. A compact man with a shaved head and hooded eyes, Bamforth spoke with a Northern English accent and carried himself with preternatural calmness, the latter of which had evidently served him well in the high-stakes world of pharma executives. While running Eli Lilly's marketing operations, he had distinguished himself by marketing one of the most successful drugs on the planet.

Cialis, the late-to-market erectile dysfunction drug, was widely assumed to be a nonstarter unable to challenge Viagra's crushing dominance. While Viagra's effects lasted four hours, Cialis's lasted thirty-six, an absurd amount of time for any but the most committed hedonists. It was up to Bamforth to sell to the world what seemed like a cut-rate product. But he saw an opportunity. While Pfizer was running ads with baseball players and NASCAR drivers to highlight Viagra's effect on masculinity, Bamforth's ads showed couples taking their time in romantic situations. It wasn't about sex: it was about intimacy, something that you couldn't simply concoct when the clock was ticking. In 2013, a decade after its launch, Cialis surpassed Viagra as the market leader, besting it with $2.2 billion in worldwide sales. It was one of Bamforth's crowning achievements.

When he arrived at UNC in October 2019, the Institute for Innovation was buzzing with activity. Amid the many research programs, committees, and funded projects he was suddenly in charge of, Bamforth kept encountering one word: *READDI*. It wasn't long before Moorman, Willson, and Edwards explained their plan to build an arsenal of broad-based antivirals to counter the threat from flaviviruses, alphaviruses, and—as per Ralph Baric's suggestion—coronaviruses. Bamforth liked the idea. It was also one of those future-facing academic programs that didn't cost very much and looked good to donors. He agreed to keep the institute's seed funding for READDI going.

This decision proved prescient. With the backdrop of the "Wuhan coronavirus" spreading across Asia and Europe, causing deaths and nightmare scenarios like thousands of cruise ships forced to quarantine at sea for weeks while the virus tore through its passengers and crew, the world was rapidly coming to grips with the power and chaos of the hidden virome. By the time the consortium's board met in February, the novel coronavirus appeared to be spreading unfettered across the globe. The board members, almost all of whom were pharmaceutical executives, listened intently as Willson and Bamforth (with an assist from Edwards) laid out their pitch for READDI, which sounded halfway between an apocalyptic vision of total viral annihilation and a utopian dreamland where medicines were made without market forces interfering. But with the threat of the coronavirus looming, Willson and Bamforth repeatedly hammered home the same undeniable truth: the novel coronavirus wasn't the first, nor would it be the last, of the viral pathogens with which humanity would have to contend. And though the notion seemed counterintuitive, the best time to plan to stop the next pandemic was while the current one was bearing down.

The consortium's board members had spent years listening to Edwards's diatribes about the myopia in the scientific marketplace. They immediately understood the stakes—perhaps better than the READDI team themselves. The problem wasn't in READDI's vision; on that, the board was aligned. Unsurprisingly, the pharma executives took issue with how the plan was going to be executed. "They basically said," Bamforth recalled, "'Guys, this is great work, but you need to double down; you need to be thinking way bigger here.'" Edwards, seeing an opening, piled on, challenging Willson and Bamforth (who was, by dint of leading the institute for innovation, ostensibly in charge of the effort) to go bold. If READDI was as important as Willson and Bamforth claimed it was, then it couldn't be some side-of-the-desk lark; they were talking about the future of humanity, after all. And yet, in the ten or so months since Moorman and Willson had "launched" READDI, they hadn't thought through much beyond the scientific rationale.

Bamforth got the message. Used to running a six-hundred-person division and marketing products worth billions, he returned from the board meeting cold-eyed and ready to amp the project up. The READDI trio—Willson, Moorman, and Baric—were adept at scientific discovery, but they were out of their depths in the marketplace, where Bamforth did his best work. Edwards often bemoaned the lack both of rigorous processes and of assembly line logistics in academia. Bamforth had spent three decades working those systems into his psyche. He established strict time lines and deliverables for READDI, set up a board and central coordination, and began to market the project to other universities and funders. In its new, muscular form, READDI was publicly launched on April 9, 2020, with the stated goal of having five novel antiviral therapies against alphaviruses, flaviviruses, and coronaviruses through safety and efficacy trials within five years. The second goal was raising $125 million to make it all happen.

A five-year window was far too long to develop a cure for COVID-19. But READDI hadn't been designed to solve the current pandemic; it was meant to protect humanity from future ones. It would ensure that whenever the next pathogenic virus flared—and it would, sooner than we would like—we wouldn't once again be caught flat-footed. In such a scenario, READDI would have identified and stockpiled the therapeutic drugs most likely to work against the largest number of viruses. Bamforth's job was to sell the idea to the world, a task reminiscent of his Cialis days: convincing people that a product that seemed ill-suited to the market could actually change their lives. And as with Cialis, people were listening. "Everybody, when you explain the issue, when you talk about the frequency of pandemics, when you talk about what we're going through right now, and you say, 'Do you think we should have a strategic plan to develop antivirals ready for the next one?' Everyone says, 'Yep, we agree.' We're like: 'Okay. That's the good news.' And then you have to get people to open up the purse."

To Edwards's delight, Moorman, Willson, Baric, and even Bamforth had begun trumpeting the kinds of statements about open science that had long been his bread and butter. SARS-CoV-2 had proven that there was no market force to develop an antiviral for a virus that hadn't emerged; and yet, if that force existed, humanity wouldn't have been caught flat-footed by the novel coronavirus. Leaving solutions to pharma alone had let looming threats fall far down the priority list. Moorman put the problem succinctly: "How much would you have paid me for a COVID-19 vaccine a year ago? The answer is nothing." But as SARS-CoV-2 spread exponentially through the first quarter of 2020, those old rationalizations became bankrupt. Moorman predicted that in the short term, the obsession with tackling COVID-19 would limit READDI's salability. "But once we get through this," he said, "we'll have a window when people will have a memory of this moment, before they go back to not worrying about pandemics anymore." READDI would have to hit that sweet spot.

And then there was Baric. As he surveyed the landscape of drug discovery around COVID-19, he saw it through the forty years he had spent seeking to understand and control coronaviruses. Everyone wanted to develop a cure that worked against SARS-CoV-2, but no one recognized that the virus and our response to it were part of a pattern. In 2003, the U.S. government paid $180 million to commercial companies to develop a vaccine for SARS; by 2004, the funding had stopped entirely, with little to show for it. Ten years later, the cycle repeated itself with MERS: the emergence of a novel coronavirus, a short frenzy of activity, and a failure to produce any actual cures. "The fundamental problem is this: you don't know what's coming," Baric said. "A company doesn't want to put money against a future virus because there's just one target: the current virus." Baric saw in READDI a way to break the mold of what he derisively described as "one drug, one bug" thinking. "READDI is the 'one drug, many bug' solution," he explained, a way to stop the entire coronavirus family in its tracks—both its ancient, long-lived ancestors and their progeny to come. The best part was, READDI wouldn't necessarily need to develop new drugs; as Baric knew, there were always diamonds in the rough waiting to find their true calling. Remdesivir was a prime example that success stories might just be lying on a shelf somewhere, collecting dust, and a good reason for pharma companies to want, in theory, to partner with READDI. "Companies," he said, "love to get lucky."

FOR HIS PART, ALED EDWARDS ADMITTED TO NOT HAVING THOUGHT much about pandemics before COVID-19. Nevertheless, he adored the fact that READDI was a world-saving initiative that also stood as a countervailing force to the private market. If the pathway to open science was to battle future coronaviruses instead of long-neglected viruses that currently tormented millions of the world's poor (like malaria, tuberculosis, and leishmaniasis), so be it. He would find a way to make it work in his favor.

READDI had even inspired Edwards to find common ground with Bamforth, the kind of pharma executive he normally loathed.

Edwards had made the case that open science was the new wave of innovation, and Bamforth, despite his deep belief in the power of pharma to cure what ailed humanity, was intrigued. And while the Structural Genomics Consortium had initially come to the University of North Carolina to expand its protein-mapping work, the pandemic, like a retreating glacier, had revealed a hidden gem: Ralph Baric. It seemed like the perfect fit. The only question was whether the three men—an iconoclast who disposed of allies without a second thought, a pharmaceutical marketing executive, and a laconic virological genius—could hold it together long enough to bring their plan to fruition and save humanity from repeating, on a global scale, the eighty-year wave of epidemics that wreaked havoc on the Northern Song dynasty throughout the eleventh century.

March 2020, Seattle, Washington

The eeriest change for Nick Mark, the Seattle-area ICU doctor, was in the sound. The constant thrum of conversation that made up the basic aural environment of hospital life had been silenced. Family visitations were banned, the people and their chatter gone. Entire parts of the hospital, like its operating rooms, were a ghost town. As Mark moved through the hospital ward, his off-white gown trailing behind him, the silence of the halls only served to emphasize the cacophony that confronted him when he walked into the ICU: the blaring of the room's negative-pressure fans, the hiss of the patients' oxygen supplies, the array of monitors beeping in frustrating asynchrony, and his own voice, too loud in his ears as, desperate to be heard, he shouted through the polyurethane-coated layers of his PPE.

The first wave of COVID-19 patients that confronted Mark in the ICU presented with an especially bad form of acute respiratory distress syndrome (ARDS) called DAD, or diffuse alveolar damage. It came on gradually, insidiously, before leading to profound hypoxia, a state in which the body's respiratory cells, turned rigid, were no longer able to absorb oxygen no matter how much was pumped into the lungs. The virus was causing people to suffocate and die.

"We are doing everything we know how to do to treat ARDS to try to keep people alive," Mark said, exasperated. But medicine couldn't simply be invented out of thin air. Despite SARS killing one in ten of those it infected and MERS killing one in three, by 2020 the scientific evidence on how to treat coronavirus infections was barren. SARS had been sent back into the wild; MERS hadn't spread sufficiently to develop clinical standards of care; and OC43, NL63, and the other benign human coronaviruses didn't send people to the hospital. Years of testing convalescent plasma, viral inhibitors, host-directed inhibitors, monoclonal antibodies, and repurposed FDA-approved drugs had proved fruitless: "[T]he efficacy of treatments for SARS-CoV and MERS-CoV infection currently remains unclear," concluded a review in *Nature*, because so much of how coronaviruses affected the immune system remained unknown. Worse, scientists had found that some of the treatments were probably harming patients.

Without a track record, Mark and clinicians like him had to improvise. At first, he and the others followed the standard textbook for people with respiratory illnesses like pneumonia. They gave them oxygen, and if that wasn't working, they intubated patients and connected them to a mechanical ventilator that would force their lungs to expand and contract. If that failed, they moved on to ECMO, or extracorporeal membrane oxygenation, an imposing series of tubes, compressed-gas cylinders, and pumps on wheels that artificially oxygenated the blood and took pressure off the lungs and heart. Mark tried neuromuscular blockers, a class of drugs that caused muscle paralysis and allowed patients to rest. He gave them inhaled prostacyclin, an anti-inflammatory drug that also helped lower blood pressure so that his patients' hearts wouldn't give out. "Some of it works," he said. Most of it didn't.

The crux of the issue wasn't even that COVID-19 symptoms were so manifold or severe. ICU patients infected with the virus just simply refused to get better. On average, ICU visits for respiratory diseases kept patients on a ventilator for three to five days, after which most recovered (though about a quarter died). In the early days of the pandemic, symptoms and survival rates for coronavirus-infected patients who ended up in the ICU were similar, but with one major difference: COVID patients were on average hooked up to a ventilator for *fifteen* days, while their total time in the ICU was reaching *three weeks or longer*—more than five times as long as patients who came in with "classic" viral respiratory illnesses like severe influenza. For a system like the ICU, which was built on short stays, the implications were disastrous. The reason, clinicians discovered, was because the virus caused a "biphasic" disease, which Mark had seen repeated over and over among his critically ill patients. After being infected with SARS-CoV-2, the human immune system mounted a response against the virus (the first phase). If that response failed, the immune system moved on to more desperate measures (the second phase). Mark was encountering patients who, after a weeklong fight, appeared stable, the virus in retreat. Suddenly, though, they would spiral into multi-organ failure, hanging over the precipice of death. The ruptures in their organs were caused by cytokine bursts, an immunologic scorched-earth strategy akin to the body's releasing an internal nuclear bomb in a last-ditch effort to kill an invader at the cost, sometimes, of the patients themselves.

A second wave brought another morbid twist. By the end of March 2020, the bodies of the newly infected weren't like those arriving from the nearby long-term-care homes, octogenarian survivors of decades-long running battles between viral pathogens and human immune systems who had filled the hospital in the first wave. These new bodies were young and healthy, and when Mark looked through his face shield and into the ICU rooms, it was like a warped mirror had been held up to his face: the patients were frontline healthcare workers in the prime of life, just like him. It was terrifying. "The psychological effect of having staff sick in the hospital where they work is just utterly devastating," he said, "because you realize it only takes one mistake for you to be like them."

Mark loathed the "healthcare workers as heroes" discourse, which he saw as insidious. "You expect heroes to die," he said, his buoyant optimism dropping away. "Once you start treating people as heroes, it almost normalizes sacrifice. And there really shouldn't be sacrifice; you shouldn't have healthcare workers dying." To him, the mass casualties weren't cause for arch ceremony, but rather a failure of the system that needed to be fixed. The media-burnished melodrama that implied that the deaths of his colleagues were an inevitable sacrifice, and even perhaps a welcome one, was too convenient. "As nice as it is to see homemade signs at the hospital and in parking lots that say, 'Thank you,' I'd much rather not be lionized and make sure that we're all safe." In the absence of effective treatments, though, the homemade signs would continue to proliferate.

Chapter 11

THE TECHNICAL TERM IS
A SHITSHOW

I N THE MONTHS AFTER IAN LIPKIN FLEW BACK TO NEW YORK FROM his under-the-radar trip to Beijing, he was struck by how rapidly the world was changing as COVID-19 outbreaks coalesced into a generalized pandemic. While there was some good—colleagues were more willing than ever before to share data and collaborate— most of it was just plain bad. New York was transformed, its usually brisk pace stopped in its tracks by stay-at-home orders that kept the streets empty and any interactions between people furtive and filled with paranoia over getting infected.

Then, events took a darker turn. "People were boycotting China-town, and there was an enormous amount of anti-Chinese senti-ment," he said. This had been spurred on by President Donald Trump's repetition of anti-Chinese tropes about the virus. It was ignorant and wrong, and Lipkin, who was familiar with how China's government would react to the hostile messages, saw it as deeply counterproductive. "All it does is it makes it more difficult to get information from China that might be useful in saving lives."

The racism had even pervaded public health protocols to control the spread of the virus, throwing them off-kilter and inadvertently placing more people at risk. Lipkin had witnessed it firsthand im-mediately upon his return from Beijing. Instead of passengers being allowed to move breezily through the airport and out to waiting taxis, Lipkin's flight was met on the gangway at the airport by officials from Homeland Security and the CDC. Everyone exiting the plane was questioned about their trip: where they had gone, why they were there, and with whom they had been in contact. While his flight was singled out, passengers arriving from Europe were allowed simply to pick up their bags and leave the airport. Even before the virus hit U.S. shores, a fog of paranoid xenophobia had already engulfed America.

Lipkin wasn't against taking proper measures to control the spread of the epidemic. Still, the exclusive targeting of flights emanating from China struck him as a product of racist and nationalist sentiment rather than a true public health response. As the months passed, Lipkin tracked the splintering of the original SARS-CoV-2 into variants, the paths of which could be used to trace outbreaks in the United States back to where they originated. Despite the grave political statements about viral threats emanating from China, it became clear to Lipkin that it was early flights from Europe, not China, that were the graver threat. "The European flights had people who were infected, who weren't stopped at all, and they were the ones who ultimately carried the virus variant that was responsible for the disease that we saw in the eastern United States and in the Midwest," he said. It was this early unforced error that allowed the COVID-19 pandemic to get a foothold in the United States.

Lipkin watched, horrified, as the racist tirades against China soon morphed into conspiracy theories claiming that the novel coronavirus was a bioengineered weapon designed at the Wuhan Institute of Virology by Shi Zhengli. Lipkin was, like anyone working on coronaviruses, deeply familiar with Shi's work and had met her a few times over the years. "Is she conscientious, is she a good scientist? Yes," he said. "Do I think that she might have deliberately done something to threaten the world? No. Is she the kind of person who did sloppy gain-of-function experiments? I don't think so," he said. "That's not her as I know her." He believed that if the theories about Shi were left unchallenged, they could derail global efforts to control the virus. Though conspiracy theories were beyond the scope of most scientists, Lipkin was uncommonly attuned to the role that politics and pervasive cultural fears played in intensifying epidemics. For the veteran epidemiologist, prejudice and paranoia weren't distractions; they were the pathways that emerging pathogens used to sicken and kill humans.

Lipkin had been commiserating with a close-knit group of virologists who had been reviewing the genome sequence of the novel coronavirus—which included the Australian virologist Eddie Holmes, who had collaborated with the Chinese team that initially released the SARS-CoV-2 genome. The group was well versed in structural biology, and when they did a deep dive on the coronavirus genome sequence, it was obvious to all of them that it had not been bioengineered. They decided to prove this to the world.

In a study published in *Nature Medicine* in early March 2020, Lipkin and his colleagues explained that the genomes of SARS and the novel coronavirus had five different amino acids in each spike's receptor-binding domain. While those differences appeared to make the new virus more transmissible among humans compared to SARS, when they compared it against the structure of all known human coronavirus receptor-binding domains, they found that these mutations were far from the best solution coronaviruses had evolved to enter human cells. If the goal was to engineer a perfect bioweapon, Lipkin pointed out, why wouldn't the evil scientists have used the best mutations the *Coronaviridae* had designed themselves?

Beyond the problem of the imperfect mutations, Lipkin sought to set the record straight on just how much virologists really knew about coronaviruses. He and Baric had been close friends for fifteen years—Lipkin was fond of teasing Baric for having strayed so far from his competitive swimming career—and had collaborated throughout that time. This gave Lipkin candid insight into what humanity really knew about what made some coronaviruses so deadly. The answer was: not that much. "Even if you decided that you wanted to make such a virus, there's no blueprint for doing it," he said. The fact was, even Ralph Baric, who knew more than anyone about coronaviruses, readily admitted that he had only the fuzziest sense of what some of the viruses' components actually did, and his approach to discovering their functions was generally to delete specific components and see how the viruses reacted. (Case in point: while Baric discovered that remdesivir and EIDD-2801 sabotaged the coronavirus-copying machine by swapping out the nucleotides cytosine and uracil for junk molecules, he didn't know how the drugs actually accomplished this at the molecular level. And if he didn't, no one did.) With genomes that stretched to thirty thousand molecular bases, it was impossible for anyone to know how a new mutation might improve viral transmissibility or virulence; there were no microscopes powerful enough to allow for a straightforward look at single nucleotides. Arming a bioweapon by creating a key mutation would have been like trying to find the right screw in a bin that held thirty thousand, all of which were roughly the same size . . . while wearing a blindfold.

While Lipkin had no doubt that SARS-CoV-2 had emerged naturally from the hidden virome, he conceded that how the novel coronavirus had made its way into human populations was still unknown. "I felt less strongly than my colleagues that we could exclude the possibilities of some inadvertent introduction through humans," he said, "but we weighted the three, ruled out a synthetic virus, and then the feeling was that it was more likely that it simply originated in nature, as most of these things do, rather than that it came from the Wuhan Institute of Virology." Lipkin hoped that his contribution would help cooler heads prevail, though he had his doubts. "Everything is everywhere: that's really the point. And until such time as people realize that there is no way to contain viruses, that they travel on birds and planes and travel and trade—you know, you're going to have this us-versus-them mentality. And that is not productive."

March 16, 2020, Seattle, Washington

Three months after the SARS-CoV-2 genome was released to the world, four healthy people in Seattle were given injections of a vaccine candidate dubbed mRNA-1273 that, at the time, felt like a futurist fantasy. They were the first among forty-five enrolled in a Phase 1 clinical trial that, with great fanfare, had been designed to test the vaccine's safety and the kinds of immune responses it might produce. The injections represented the culmination of a decade of work that began when Derrick Rossi watched pulsing light emanate from within a mouse's thigh and continued through Tal Zaks's assiduous cultivation of an NIH-partnered vaccine program at Moderna. This placed the trial participants at the tip of the spear of mRNA technology, though it was not yet clear whether they were in its path or riding it to victory. That's what the Phase 1 trial was meant to establish.

Though Zaks exuded confidence in mRNA, the number of vaccines using the technology that had been successfully brought to market was exactly zero. This had led to doubts among many notable vaccinologists, Bob Brunham included, who did not understand the sudden frenzy around using experimental platforms when there were already so many safe and reliable ways to make vaccines. Tal Zaks saw things differently: the existing vaccine platforms were woefully ill-equipped to adapt to emerging threats. Spurred on by SARS-CoV-2, the decade-long battle between vaccine classicists like Brunham and vaccine modernists like Zaks finally came to a head.

When the target of the vaccine was a virus that posed a threat to the entire human population, open questions about how the vaccine might work became profound sources of anxiety. Even Zaks had his moments. Shortly after the day Moderna began producing mRNA-1273, he woke up in a cold sweat, suddenly anxious that the genomic sequence was faulty and that all might be for naught. It was a fleeting paranoid fantasy, but it underscored just how high the stakes he faced were. Since 2015, Zaks had helped Moderna develop multiple mRNA vaccine candidates. But without a track record of successful vaccines, without data showing that mRNA-1273 was safe in animals, let alone in humans, and with the fixed and despairing glare of the entire world upon them, it was far from certain that Moderna would, as Zaks had long ago boasted, save the world. The company had been in the pharma market for less than a tenth of the time of most of its major competitors, many of which could trace their founding back to the nineteenth century, and the potential threats to their continued existence were many. If the technology didn't work, it might spell the death knell for mRNA vaccines. Even if their vaccine worked but they were able to bring it to market only long after their competitors, doubts about the viability of mRNA vaccines (heralded as a rapidly produced, easily programmable technology) would seriously undercut their credibility. Without data, and fast, Zaks knew that Moderna's COVID-19 moon shot might just end up sinking him.

Despite vaccines being heralded as the way to end the pandemic, Bob Brunham was skeptical that any of the experimental candidates would end up producing a viable cure. "When a SARS-CoV-2 vaccine comes on the scene, it's going to be prime time, and we have to be confident that this vaccine does no harm and that it works," he said. "We really need to play this one safe."

Brunham watched from afar as Moderna and Pfizer launched their mRNA vaccine trials, his antipathy evident. "None are going to give us an early win," he said. Not privy to the inner workings of what Baric, Graham, and Zaks had been doing, Brunham was convinced that the time line to getting experimental vaccines off the ground would still be measured in years, not months, though he conceded that with enough resources, that lag time might shift. "As a scientist, I love these approaches," he said, "but cash can never compensate for a flawed idea. Moderna has hundreds of millions of dollars; but if it doesn't work, it doesn't work."

Brunham had come by his concerns honestly, though it was hard to believe that, deep down, his dire warnings about experimental vaccines weren't at least partly a result of his feeling left out of the game. After all, he had been working on vaccines for four decades, but when a coronavirus fit enough to require one finally emerged, he had nevertheless found himself on the outside looking in. As he surveyed the literature on vaccines, the wildest ideas seemed to be gaining the most traction at the expense of the proven methods. There was the biotech firm Inovio, which wanted to use brute force (in the form of a three-pronged electrified needle) to open cell walls and let its DNA vaccine enter. There was Symvivo, which proposed cultivating synthetic bacteria to grow DNA vaccines inside a person's gut. It was as if the whole world were about to jump off a cliff when there was a perfectly safe trail—there, in plain sight—that could lead us all down the mountain without injury.

So, he went to work. "In the face of a pandemic where time is of the essence," he wrote in a March 2020 op-ed, "resources for established vaccine technologies should be prioritized." What the world needed were vaccines that had been fired in the blistering forges of scientific peer review. With millions of lives hanging in the balance, this was no time for games of chance.

Brunham—the first to sequence the SARS genome, the scientist who had launched SAVI and tested multiple SARS vaccines in animal models—figured that his informed opinion would hold at least some weight. He was wrong. Not a single newspaper would publish his op-ed warning of the dangers of blindly rushing toward experimental vaccines. Nobody cared, it seemed, what senior scientists thought anymore. Right at the moment when his wisdom, collected over a lifetime of experimentation, was more valuable than ever, Brunham had lost his voice.

His options to make an impact dwindling and his scientific skills no longer needed, he found himself increasingly exasperated. While his lab was still equipped to test out vaccines, the sheer size and deep pockets of the high-profile trials run by Moderna, Pfizer, and Astra-Zeneca scared funders away. Why bother funding something that, by the time it was launched, would be running far behind its competitors? In the wake of diminishing prospects, and in the twilight of his career but with a scientific mind as powerful as ever, Brunham found himself reduced to sending messages to generic government email accounts in increasingly vain attempts to call attention to the dangers of an errant vaccine race.

SARS had been a transient threat, too fragile to serve as the perfect enemy. If it had shown a little bit more fitness, Brunham knew that SAVI might have taken off and saved the world. When SARS-2 finally hit, the destruction wrought by the sublimely effective pathogen was a challenge magnitudes greater than the one with which he had first contended. Brunham, it seemed, was destined to live out this crisis from afar, mutely watching as the virus tore a destructive path across an ill-prepared world.

BARNEY GRAHAM HAD SET JULY 1 AS THE DATE TO LAUNCH PHASE 3 human trials for Moderna's mRNA vaccine. Anything later, and getting a vaccine into arms by the end of 2020 became a remote possibility. In the months leading up to the deadline, Baric had been busy playing his part to give Graham what he needed. As soon as Moderna put its vaccine in production on January 13, 2020, Graham had tasked Baric with collecting data showing that it was safe and effective in mice. Though vaccine trials usually moved sequentially, with human trials starting only after animal trials were complete, the pandemic's rolling spread across the planet required a different tack. Graham was running the small Phase 1 trial for Moderna's vaccine at the same time Baric ran safety experiments in mice. Graham's dual-track strategy would have been reckless had he, Zaks, and Baric not already tested an mRNA vaccine against MERS, which Baric had shown was both effective and safe. Still, the pressure was on Baric to lay the groundwork for Phase 2 human trials: if he failed to detect a potential vaccine-related hazard, he would be placing the thirty thousand trial participants in danger.

Meanwhile, the numbers of the pandemic dead were rising precipitously, by the tens of thousands, each day. In the second half of March—less than three months since the first case on U.S. soil— more than a hundred thousand Americans had become infected with COVID-19 and hundreds were dying each day; globally, almost eight hundred thousand people had become infected. In the bubble of his laboratory, Baric sought to retain focus on the task at hand, but it was impossible not to let the pressure seep in: the deaths would continue to mount until a vaccine arrived, and one of the most promising, Moderna's mRNA-1273 candidate, was waiting on him to be tested in human trials.

He wasted no time, setting up mouse challenge studies, whereby the animals were infected with a mouse-adapted version of SARS-CoV-2 and then inoculated against it with mRNA-1273. While he could move fluidly through many of the steps of the experiment, there were some facets of his work that simply could not be rushed, no matter the state of the pandemic out in the real world. Baric and his team had to observe the mice over a five-week span, to determine how well the vaccine protected the tiny creatures, how long immunity to SARS-CoV-2 might last, and whether any adverse reactions emerged. Only then could they analyze the data and send it back to Graham to be fed into the larger mission of getting the vaccine into the human population. Waiting was part of the job; Baric knew this and had throughout his career found solace in the complex polyrhythms of laboratory work. This time, though, with the world plunging deeper into a nightmare that he had been presaging since MERS, there was no joy to muster amid the silence.

By March 2020, the scientific marketplace for COVID-19 cures had become painfully crowded. Coronavirus science was no longer the backwater discipline in which Ralph Baric had happily spent his career working. For every Moderna—with a rarefied technology ten years in the making, a market capitalization in the billions, and a Phase 1 trial in operation—there were the wishful thinkers who saw in COVID-19 an opportunity to flout their wares. It was hard to tell the real science from the many dubious claims that circulated amid the general frenzy around COVID-19. More than five thousand peer-reviewed studies about COVID-19 had already been published. The discipline's heat threatened to make it the world's most absurd, as anybody with even the vaguest idea for stopping the spread of the virus could be a pretender to the throne.

To get in the game, all you had to do was register a clinical trial. This didn't, of course, mean that you actually had to have the funding or any real plans to carry one out, so there was no downside to throwing your hat into the ring. Maybe, just maybe, some deep-pocketed investor would read your trial title and figure that it was worth throwing money at. Or you could use a trial registration to pad your academic résumé in the hope that the reflected glow of the actual science being done on COVID-19 would make you seem more impressive. For those scientists working on topics entirely unrelated to the pandemic, it became obvious that simply adding the words "During the COVID-19 Pandemic" to your study title made your work much more likely to be funded. And so, naturally, registrations popped up for clinical trials to study all kinds of COVID-19 cures: mindful eating, shadowboxing, a six-minute walk, cannabis, umbilical cells, acupuncture, menstrual blood, vitamin C, vitamin D, herbal supplements, soybean water, stem cells, and even Thalidomide.

Amid these questionable scientific quests, some trials stood out as reasons for hope. Scientists registered studies to compare different COVID-19 tests, oral versus nasal swabs, to see which ones more reliably identified the virus, based on the fact that coronaviruses colonized multiple parts of the respiratory system (the nose, throat, and lungs) in different ways. Antiviral and steroid trials were launched to test whether their well-known effectiveness in preventing infection and inflammation would extend to COVID-19. Multiple pharma giants and biotech companies were gearing up to test whether cocktails of monoclonal antibodies (cloned white blood cells that had been developed to combat SARS but that Baric had found didn't work against near-identical SARS-like bat coronaviruses) could stop COVID-19 disease progression. All of the activity provided hope that Nick Mark and clinicians like him would soon have real therapies with which to treat their patients.

With confirmed cases still well under a million across the globe, and the science on SARS-CoV-2 mostly speculative, the cure might be found anywhere. Still, the line between the ridiculous, the self-serving, and the true contenders was blurry. For some, this early flood of scientific data was overwhelming. For others, being caught up in the rising tide was both exhilarating and potentially lucrative. Then there were events spurred on by forces beyond money and science. And on March 31, 2020, one of the more bizarre cures proposed in the course of the pandemic made its introduction. On that day, a conspicuously large cluster of trials—six in all—were registered to study the impact of an until recently obscure drug called hydroxychloroquine.

Hydroxychloroquine, an antimalarial treatment that had been repurposed to treat chronic autoimmune diseases like lupus, was an odd place for the forces of politics and science to clash. It started innocuously enough: On March 16, a small French study reported that people infected with SARS-CoV-2 and given hydroxychloroquine (along with the antibiotic azithromycin) had less severe COVID-19 disease. While the results were interesting, there was much to question about the study, including its size (only forty-two people were enrolled), its methods (for unexplained reasons, data from six participants were removed), and its remarkably swift pathway to publication (it was submitted to the *International Journal of Microbial Agents* on March 16, ostensibly peer-reviewed that very day, and published the next day), which then led to charges of conflict of interest (the first author of the study also happened to be the founding editor of the journal that published it). Prior to the pandemic, a study that small, and one that raised that many red flags, would have been quickly forgotten in the ever-forward rush of scientific discovery, like a small eddy momentarily whirling in a river. But fate intervened.

Two days after the French study was published, a voice carrying considerable authority, none of it scientific, weighed in. Hydroxychloroquine was a "game changer," President Donald Trump announced at a White House press conference on March 19, 2020, the muted sound of camera shutters whirring as he spoke to the assembled reporters. Referring obliquely to the French study, Trump extolled the drug's virtues and, in typical fashion, wildly overstated both its effectiveness and its availability. "It's shown very encouraging—very, very encouraging early results," he told reporters. "And we're going to be able to make that drug available almost immediately." It was a grandiose claim at a time when uncertainty about the novel coronavirus—its transmissibility, the role its various parts played in eliciting disease, its capacity to kill—had fomented a desperate desire for hope.

The president's full-throated endorsement of an unsanctioned drug was meant to salve an increasingly terrified public, but it was also the worst-case scenario for Davey Smith,* the head of UC San Diego's Division of Infectious Diseases and Global Public Health. A medical doctor and virologist, Smith had, like so many clinician-scientists of his generation, cut his teeth seeking cures for HIV. He had had a front-row seat across a period of three decades as a global push to end the AIDS pandemic turned infection with HIV, that master of mutation and escape, into a chronic condition not unlike diabetes or high blood pressure (for those who had access to HIV antiretroviral drugs, of course). Those victories, albeit partial and hard-won, were still cause for celebration, though Smith was more inclined to focus on what had gone wrong. There were the three decades of failure in developing both HIV therapeutics and vaccines, despite the world's greatest minds and billions of dollars invested in the effort. More galling, Smith had borne witness to what happened when access to lifesaving medicines was brutally denied because of political indecision or spite. And he had seen how far people at the edge of hope would go, dosing themselves with useless prophylactics or even toxic chemicals in the search for so-called cures, to save themselves from the incurable virus. It was an inevitable part of the HIV pandemic. It was only a matter of time, Smith knew, before COVID anxiety, rippling across the totality of the

* Disclosure: Davey Smith and I are colleagues at UC San Diego.

world's almost eight billion humans at risk of infection, unleashed the very same forces with renewed power.

Behind his amiable smile and affability, Smith, who grew up on the outskirts of Chattanooga, Tennessee, carried a certain Southern gothic sensibility upon which he relied to carry him through the failures and frustrations he encountered. A literature major turned physician turned infectious disease scientist, Smith spoke in equally rapturous tones about the works of Flannery O'Connor as he did the mechanics of DNA transcription. Like many of the clinician-researchers of his generation who had devoted their careers to battling the HIV/AIDS pandemic, Smith had been touched by the traumas and failures of the global response to HIV. When asked by a reporter to expound upon the fact that COVID-19 had introduced new fractures into society, Smith became incensed. "Nothing about this is new," he shouted. "This is how it always happens." The people who were dying of COVID-19, he pointed out, were the same type of people who had been dying of AIDS and every other epidemic. They were overwhelmingly poor and marginalized and disproportionately people of color. The only thing that had changed was the pathogen.

Smith, early on in the pandemic, had a good sense of how the anxiety around treating COVID-19 might manifest. In January 2020, when the novel coronavirus was still essentially a Chinese domestic problem, he had sent a proposal to the U.S. National Institutes of Health for a large-scale randomized clinical trial to test the effect of hydroxychloroquine on COVID-19 disease. Hydroxychloroquine occupied a sweet spot that made it, Smith figured, potentially attractive: it had been in use for almost a century (it was put on the market in 1934); it had proven to be effective against lots of different conditions, including malaria and arthritis; and it was basically safe to use. What's more, there were a handful of laboratory experiments dating back to the early 2000s (conducted by, among others, Leen Vijgen, the Dutch virologist who discovered NL63) that showed that hydroxychloroquine stopped human coronaviruses, including SARS, from replicating. The fact that the experiments were done using mouse cells in a petri dish was, Smith knew, irrelevant to how they would be interpreted amid the COVID-19 pandemic, when faint promises took on the power of miracle cures.

Smith was as ignorant as anyone about whether hydroxychloroquine would stop SARS-CoV-2 from replicating, and he admitted to having no illusions about its being a panacea. (Baric, for his part, had never tested the drug as a potential coronavirus therapy.) Still, Smith knew that if there were even an indication that it might be curative, and if the drug were available, people would start using it. The evidence on hydroxychloroquine was pretty weak, but clinicians the world over, desperate to help their patients by any means necessary, were almost certainly going to start prescribing it because it was already so widely available. A high-quality trial of hydroxychloroquine was needed, Smith implored, before people took matters into their own hands.

Smith's passionate entreaties failed to register. The NIH just blew him off. Undeterred, he sent the same proposal to the U.S. Department of Veterans Affairs, one of the country's major funders of virological research. For good measure, he also forwarded it in January to the NIH-funded AIDS Clinical Trials Group, perhaps the most august assemblage of virologists in the world, who had begun shifting their considerable resources from the HIV pandemic and toward COVID-19. The response? "Crickets," said Smith. Despite the surge in research funds directed toward the pandemic by January, nobody was interested in funding trials of a drug that reasonable scientists believed wouldn't work. Nobody funding research, that is: on the ground, the situation was starkly different. Borne initially by a weak trail of outdated data from in vitro experiments with mouse cells, hydroxychloroquine was increasingly listed by clinicians in case reports as a treatment for those felled by the novel coronavirus.

Nick Mark had seen firsthand how clinicians rushed to test new and unproven treatments on patients, even at the risk of potentially harming them; SARS-CoV-2 was just the latest and most extreme example of this phenomenon. "It's very easy for things that are unclear in scientific writing to lead to public misperceptions," he explained, "which has led a lot of people to look for these magic, silver bullet–type cures." It was an understandable urge for clinicians facing a new enemy with outdated weapons and helplessly watching their patients die. "For most of us who work in the ICU," Mark said, "we're pretty skeptical of the magic cures. But I think part of the reason why people cling to things like hydroxychloroquine is because they had a sense that, you know, nothing that we were doing worked."

By mid-March, this widespread clinical despair had caused a surge in doctors all across the world prescribing the drug for patients sick with COVID-19. This spike in hydroxychloroquine use then spurred a wave of small, methodologically suspect studies from China, Korea, the United States, and, finally, France. As those papers reported promising preliminary results, clinicians redoubled their efforts to use the drug, thereby completing a feedback loop that saw anecdotal evidence fuel clinical prescribing, which fed into even more poorly designed studies with dubious results, which then cemented hydroxychloroquine as a key therapeutic in the fight against COVID-19. It was exactly what Smith had predicted would happen way back in January.

And then the president stepped in. At Trump's March press conference, his seemingly off-the-cuff endorsement of hydroxychloroquine (which most people had never heard of) as a "game changer" against SARS-CoV-2 turned it into front-page news. Overnight, everyone had an opinion on the drug, many of which had nothing to do with scientific evidence about coronaviruses and everything to do with their opinion of the man who had just become the drug's promoter in chief.

Throughout the early months of the pandemic, scientific misinformation had been spreading as quickly as the virus. Trump's elevation of hydroxychloroquine tipped the scale, ratcheting up the reproductive ratio of false claims by magnitudes. The week following his March 19 press conference, more than one hundred thousand posts about hydroxychloroquine were shared on Facebook. Within ten days, under pressure from the White House, the U.S. Food and Drug Administration announced an emergency-use authorization for the drug, and millions of doses were distributed to hospitals despite the complete absence of high-quality evidence. (This, in turn, caused severe shortages for people who relied on daily hydroxychloroquine prescriptions to control serious chronic illnesses like lupus and rheumatoid arthritis.) "What do you have to lose?" Trump retorted when questioned about the flimsy data a few days later. "I'll say it again: What do you have to lose? Take it."

Surveying the chaos, Davey Smith offered his objective diagnosis. "The technical term," he said caustically, "is a *shitshow*." The worst part was that the disturbing convergence of political power and scientific ignorance could have been prevented with an early trial testing to see if there was any merit to the claims about hydroxychloroquine.

But that ship had long sailed. After the president's remarks, though, others finally came to agree with Smith. After the initial January blow-off, the NIH-funded AIDS Clinical Trials Group came back to him, shaken, it seemed, by recent events and desperate for Smith to start the study he had proposed three months earlier. Smith, who kept a folder of all the rejections he received as a teaching tool for his students, was typically sanguine at the news. "The frustration is that we should have done it in January," he said, "but denial is the first human defense mechanism."

The sudden urgency to get the trial off the ground and the blazing speed of science during COVID-19 partially made up for the months of delays. For those who had followed the tributaries of coronavirus science prior to the pandemic, it was also shocking to see how differently coronavirus research was now treated: after all, the NIH moratorium on gain-of-function experiments was lifted only in 2017, after three long years during which no new experiments using pathogenic coronaviruses could be run.

After getting okayed in the last week of April, Smith's study was officially launched on May 1. By May 7, less than a week later, he had enrolled his first participant. "It was the fastest trial that's ever been started at the NIH," Smith said. "These study protocols typically take over a year to develop and six months to launch. We did it all in a few weeks." He was evidently proud of the effort, despite the political machinations that had led to its launch. He would soon learn, though, that his stint in politics was far from finished.

OVER THE COURSE OF RALPH BARIC'S CAREER, HIS VISION HAD GROWN in lockstep with the threat that coronaviruses posed to humanity. He had entered the discipline to find out how coronavirus genomes had bucked the laws of nature to become so absurdly large, a question delightfully removed from the real world. But SARS, with its shifting variants, had forced Baric to confront how ubiquitous, variegated, and deadly the family really was. MERS had proven that SARS wasn't a rare event, but rather the start of a pattern of accelerating zoonotic jumps. Coronaviruses were the bridge that forcefully reminded us how much of the mechanics of our bodies we still shared with our animal brethren. While Baric was shocked but not surprised by SARS, by 2018 he was wholly resigned to the next imminent threat (a human coronavirus much fitter than the poorly adapted SARS virus, which would make it much deadlier) and in the unenviable position of being one of the few people in the world with the skills to do something about it.

It didn't take long for Baric's years of obsessive experimentation to pay off. On May 1, 2020, on the back of human trial data informed by the experiments he and Denison had begun in 2013, Gilead's request to have the FDA provide an emergency-use authorization for remdesivir was approved, making it the first therapy on the market to treat COVID-19. The clinical trial, which was run by the National Institute of Allergy and Infectious Diseases and did not involve Baric, found that COVID-19 patients given remdesivir recovered, on average, roughly 30 percent faster than those who weren't given the therapy, which meant that sick people would leave hospitals faster, making more ICU beds available for the newly infected, and that, ultimately, lives would be saved. The antiviral with the power to insert itself into replicating coronaviruses, hide from the proofreader, and produce error catastrophe was exactly what Baric had predicted it would be: a potent weapon against a novel pathogen. Announcing the results, Anthony Fauci described the findings as far from a "knockout," but still a "very important proof of concept." In short: it would help them win some battles, though it wasn't enough to win the war.

It was a strangely anticlimactic victory for Baric. Since 2013, he and Mark Denison had been testing remdesivir to see if it was the answer to their "one drug, many bugs" problem. And when a new pathogenic coronavirus emerged, as Baric predicted it would, remdesivir proved its worth: it worked. And yet, when the FDA approved it and it finally came to market, the world seemed largely unmoved. To be sure, May 2020 was a time when biotech companies and pharma giants the world over were making hyperbolic claims about their own purported treatments, promising the stars and so far failing to deliver anything but a dark and moonless night. Baric, instinctively self-effacing, was loath to talk up the results of his work, preferring to let it speak for itself. And yet, after seven long years of sifting through hundreds of thousands of drugs, he and Denison had done what the others could only dream of: they had quietly fired humanity's first successful shot against SARS-CoV-2.

Baric wasn't much for reveling, though, and he admitted to some skepticism about remdesivir as a pandemic holy grail. True to form, rather than basking in the satisfaction of seeing the drug approved, he focused instead on bolstering humanity against its possible failure. And this meant turning back to the only other broad-based antiviral, EIDD-2801, that he and Denison had identified amid the hundreds of thousands they had investigated since 2015. As soon as the SARS-CoV-2 genome was released in January, Baric (while simultaneously testing Moderna's vaccine candidate on mice) began testing the effects of EIDD-2801, which was also known as molnupiravir, on the novel coronavirus. Ensconced in his BSL-3 lab, in which *Mus musculus,* the humble laboratory mouse, far outnumbered its human masters and wherein the full range of the coronavirus family resided in a neatly ordered catalogue, Baric engaged in his own version of recombination. His coronavirus panel, like an extended family portrait, was now at sixteen different strains and many more variants, spanning the gamut between bat, camel, rodent, bird, pig, and human strains. The roster included all the known human coronaviruses, the laboratory stalwart MHV (mouse coronavirus), and bat coronaviruses unearthed in the caves and mine shafts of Yunnan Province by Shi Zhengli. These represented the four corners of the family: the alpha group (which includes NL63), beta group (SARS, MERS, SARS-CoV-2, and others), delta group (primarily swine and bird CoVs), and gamma group (dolphin and whale CoVs, and more).

Remdesivir and EIDD-2801 were the twin jewels in the crown of Baric and Denison's prolonged search for weapons against future coronaviruses. Remdesivir was the first of the two that was approved, but this was a side effect of its erstwhile loser status: Gilead had run it through so many different trials in a futile attempt to find a use for it that the company had, at least, already shown the drug to be safe for use among humans.

Less was known about EIDD-2801. George Painter, the director of the Emory Institute for Drug Development, which had developed the drug, had passed it off to Baric and Denison shortly after developing it. In the years since, EIDD-2801 had been acquired from Painter's group at Emory by Ridgeback Biotherapeutics, which had then partnered with the pharma giant Merck to develop it further. One of the driving forces behind the drug's upward path through the marketplace had been Baric's rigorous testing of its performance against his panel of coronaviruses in different environments. But there had been no human trials of EIDD-2801; the only published studies had been run in ferrets, to see whether the drug blocked the spread of influenza, which seemed, prior to COVID-19, like a more practical application for it than coronaviruses.

When SARS-CoV-2 emerged, EIDD-2801 was largely forgotten amid the excitement over remdesivir and the speculation about other market-ready (but by no means proven) therapies like hydroxychloroquine. Still, Baric had long since believed that the drug might rival or even surpass remdesivir as a pan-coronavirus treatment in the real world. The key reason stemmed from how the two drugs were taken. Remdesivir required temperature-controlled storage and was available only by injection, because it couldn't survive passing through the rigors of the human gastrointestinal tract. Not so with EIDD-2801: the drug was orally administered, meaning that it could be delivered to practically anyone, anywhere, and it was stable for weeks without being stored in a freezer. During a pandemic, when outbreaks could happen unpredictably, being able to administer antivirals quickly to people who became infected was critical, not only to saving their lives but to cutting transmission cycles from one infected person to another. This advantage positioned EIDD-2801 to be the potential "knockout" that Fauci and others were looking for with remdesivir but hadn't found. It just needed to work.

Baric's many COVID-era experiments were beyond the capacity of his lab, and he resorted to cannibalizing infrastructure from his many supportive colleagues at UNC just to keep his multiple projects afloat. They were more than happy to oblige: alongside Mark Denison, Baric was the person best equipped in the world to figure out whether EIDD-2801 could well and truly inflect the arc of the battle against the COVID-19 pandemic and alter the course of the age-old war between humanity and the *Coronaviridae.*

After seven years of searching for the "one drug, many bugs" solution, Baric and Denison went all out in testing EIDD-2801. Along with SARS-CoV-2, they ran experiments on its efficacy against SARS, MERS, and the SARS-like bat coronavirus SHC014 that Shi Zhengli had discovered six years earlier, to fully explore the drug's potential as a pan-coronavirus killer.

The "test" itself was actually a series of challenges, with the various coronaviruses introduced into different cell cultures from different parts of the respiratory system: human airway and lung cells, African green monkey kidney cells (which Ali Zaki had used to ship the sample of MERS taken from the Bisha businessman to Rotterdam), and mouse brain cells. This motley mix had, after trial and error, become the gold standard for growing and testing coronaviruses. And then, of course, there was the phalanx of mice that would serve as the ultimate arbiter of whether EIDD-2801 truly was cause for hope and human trials. From January through March 2020, Baric, Denison, and their respective teams worked tirelessly to reveal whether EIDD-2801 could be added to humanity's arsenal. As he did in his most rarefied experiments, Baric was engaged in a complicated methodological dance, mixing and matching virus, cell lines, and mice and triangulating data points to discover the truth from as many different angles as possible.

For all the complexity of the cascading set of experiments, the results that Baric published in early April were eerily uniform—and encouraging. No matter what pathogenic coronavirus or cell lines they used, when they added escalating doses of EIDD-2801 to the mix, the exact same thing happened every time: the viral load (the number of virions in a given quantity of liquid) dropped precipitously, meaning that the drug stopped the viruses from replicating. And just as Baric had predicted, it did so by fooling the proofreader. Normally, the error rate for coronaviruses was only 0.01 per 10,000 nucleotide bases; when Baric added EIDD-2801 into the mix, it increased the error rate by an astounding 800 percent. The graphs presented in the final published manuscript were blissfully identical arcs, each showing the viral load plunging after contact with EIDD-2801.

Finally, when Baric turned to his constant laboratory companions, mice, they revealed more good news, but with a caveat. Results from coronavirus-infected mice given EIDD-2801 within twelve hours of infection revealed that the drug was remarkably effective at preventing disease onset. Within a week, untreated mice lost over 20 percent of their body weight, saw rapid declines in their pulmonary function, and experienced extreme lung hemorrhages, while the mice treated early were fully protected. This was the good news. The bad news was that among infected mice forced to wait twenty-four hours to receive EIDD-2801, the drug became practically useless. For EIDD-2801 to stop a pandemic from spreading, it had to be delivered fast.

Published within weeks of the FDA's authorization of remdesivir, the results of the experiments were a sign that the battle lines might just be shifting in favor of our species. Still, Baric was in no mood to claim victory. He had proven that coronaviruses withered under the antiviral onslaught wrought by EIDD-2801. But that was only the first critical step in the drug's long trajectory to market.

The next step was to test it in human trials. Merck, the company that had licensed EIDD-2801 from Ridgeback Biotherapeutics, was figuring out its plans to run a large Phase 2/3 clinical trial involving thousands of participants who would be given the drug. This was a major undertaking at the best of times, but setting up the trial became more complicated by hesitation at Merck, the company that effectively owned the drug, around how best to structure it. As the weeks passed with Merck's study still on hold, Baric and a group of scientists decided to launch their own trial, a much smaller affair involving two hundred people, which was designed only to detect one outcome: whether the drug stopped the virus from replicating. On May 28, 2020, the first human participants in the small EIDD-2801 trial were enrolled. It would take until the following year for the results to come in, Baric and the other scientists had calculated, but at least they would be moving the science forward while Merck dragged its feet.

Chapter 12

YOU DON'T SEE MUCH
WITH A SICK MINK

June 2020, North Brabant, The Netherlands

MARION KOOPMANS WAS PATROLLING THE GROUNDS OF TWO mink farms in the Dutch province of North Brabant, surveying the scene for evidence of an escalating multispecies outbreak. Koopmans, the veterinarian and virologist who first identified camels as a key reservoir for the MERS coronavirus, had been called to the scene to investigate a new and disturbing incident amid the COVID-19 pandemic. It began at the end of March, when two workers at two different farms became ill with COVID-like symptoms, leaving one in the hospital. Two weeks later, minks at both farms began to get sick and die at an alarming rate, exhibiting the subtle but telltale signs of respiratory illness. Worryingly, the farms, which collectively housed more than twenty thousand minks, had no obvious connections: they didn't share workers, vehicles, or animal transports, making the cause of the simultaneous outbreaks a mystery. Veterinary scientists were called in. As they arrived, farmworkers presented them with dozens of minks that had died from the outbreak, their small bodies prone and light, the softness of their expensive fur still palpable through latex gloves. The vets submitted mink tissue samples to a battery of tests—influenza, hepatitis A, the bacterium *E. coli*, and other common pathogens—for which they all tested negative. But without exception, the corpses were riddled with SARS-CoV-2. Curiously, there seemed no systematic pattern in which minks got sick or died, though the outbreak had spread in a patchwork across the entire herd. But how?

Koopmans had an idea. After arriving at a farm, she walked down the long rows of small wire cages, set low to the ground under the yawning roof of the open-air barn. Within each cage sat a solitary mink, its enclosure barely larger than the occupant's slender body. A thin wooden wall separated each of the soft furry animals from its neighbors, and the cage floors were laid with a bedding of straw and sawdust. A feeding trough ran between the two rows of cages, which was continually filled with a pastelike meat slurry, bits of which fell to the ground, where opportunistic feral cats lay in wait for an easy meal. The open-air barns were designed to let fresh air filter through the cages. Still, with all the hay, sawdust, particles of meat, the moisture on the breath of tens of thousands of minks, and their accumulating feces, the smell was overpowering. It was a fetid stench made far worse by the pungent odor emitted from the scent glands of the minks, which erupted when the animals were under stress. Koopmans surveyed the caged animals; some of the minks appeared to stand at attention, raising themselves on their hind legs, gripping the wire cage with their tiny claws, and leaning forward in a gesture of suspicion or supplication. Others, curled up for warmth with eyes barely open, hardly moved at all, overtaken with a deep lassitude, which made them double captives of their circumstances.

As Koopmans continued her systematic walk down the length of the barn, she leaned in closer, looking for clues in the minks' downy gray-white fur, the source of their imprisonment and the centerpiece of a global trade worth upward of $24 billion annually. "You don't see much with a sick mink," she explained. "It's the same with many of these semi-wild animals: when you see that they're sick, then they're already near dead." The only telltale signs were a mink's eating or drinking less, a watery discharge flowing from its nose, or ruffled fur; rarely did the ferret-like mammals exhibit obvious respiratory distress. With so little to go on, this made the work of isolating sick minks difficult; in mink farms, which often hold tens of thousands in tiny cages side by side, it was next to impossible. As Koopmans surveyed mink after mink, some with ruffled fur, some with eyes barely open, some with discharge running freely from their noses, she knew that it was already too late: the entire farm was likely lost.

Over a span of two weeks, Koopmans and her team sampled the dust particles lazily swirling around the barns at both farms, using small portable air pumps. In every area they tested, viral RNA was found circulating. They then coaxed oral and blood samples from twenty-four of the feral cats that hung around for meals; seven of them had antibodies for SARS-CoV-2.

Pulling the clues together, Koopmans and her team sketched out the time line and causal path of the outbreak. It was, they concluded, a "reverse zoonosis." The two human workers initially infected with SARS-CoV-2 had introduced the virus into the mink herd, within which it then spread freely. Koopmans wasn't surprised. She had been studying coronaviruses in animals since 1992, when she analyzed the blood of U.S. dairy cattle afflicted with winter dysentery, and she knew how quickly members of the *Coronaviridae* could overtake a herd. As a veterinarian, she was also intimately familiar with the immunology of minks and their close relatives. "They are very susceptible to coronaviruses," she explained, "and there's virtually no barrier for full takeoff of the virus once it enters a mink farm. It's like wildfire; it really is a very explosive-spreading virus."

Koopmans was concerned about the millions of minks across the Netherlands at risk for COVID-19. But she was also worried about what the rapid explosion of cases in minks portended for the course of the COVID-19 pandemic in humans. "In the outbreaks we looked at—during those massive transmission events—the virus rapidly picks up mutations." Chance mutations that arose as the virus spread among the minks could have allowed it to transmit itself more efficiently; the same mutations could have also produced a SARS-CoV-2 variant more dangerous to humans. No one, she admitted, was yet sure which way things would go.

What she did know, though, was that the mass transmission events among minks at the two farms wasn't coincidence. Koopmans was a close colleague of the microbiologist Ron Fouchier; both had been appointed at Erasmus University in Rotterdam, where Koopmans had worked on Fouchier's controversial gain-of-function studies infecting ferrets with avian flu. Fouchier had used ferrets in those experiments precisely because of the remarkable similarities between their immune systems and those of humans. Minks, a species of mustelids closely related to ferrets, were also perfect stand-ins for human immunology. The main reason scientists preferred working with ferrets was that minks were the more savage of the two.

Koopmans worried that what she had seen with MERS in camels would repeat itself with SARS-CoV-2 in minks: the virus becoming endemic in semi-domesticated animals. If this happened, mink farms might become the Qatari slaughterhouses of the COVID-19 pandemic—the place where animal, human, and environmental health merged at the morbid intersection of executions, aerosols, and epidemics. While the differences in camel and human immune systems had kept MERS from rapidly spreading among humans, there was little stopping minks and humans from transferring SARS-CoV-2 strains back and forth endlessly, given how similar their respiratory systems were. This cycle had already begun, Koopmans surmised, meaning that it was already too late to stop without taking the most brutal action.

Within a few months of the first North Brabant outbreak, mink farms across the Netherlands were reporting sick minks numbering in the tens of thousands. The Dutch government, which had long supported the country's mink fur trade, was forced to relent. Late spring was when the mink's annual breeding period would hit its peak. Minks give birth to up to eight kits per litter, and if true to form, the mink population was about to surge up to sixfold in size, introducing a whole new generation to the deadly virus.

On June 6, 2020, the Dutch government announced the beginning of a nationwide cull. The country, which produced four million minks per year, instructed its farmers that the year's trade was over. Across the low-lying flatlands, next to the open-air barns hosting thousands of the inquisitive, clever animals, farmworkers dug mass graves. The sleek little animals were unceremoniously dumped into sealed containers, which were then pumped full of carbon monoxide gas. The minks, some of them only a few weeks old, pressed themselves against the walls, choking until they died. None, it turned out, would survive the outbreak. Nor would the Dutch mink industry: all farms, it was decreed, would permanently close by March 2021.

FOR ALL THE BALEFUL NOVELTY OF COVID-19 FOR ITS HUMAN HOSTS, the virus driving the pandemic was playing to an ancient script. For some, like Ralph Baric, it wasn't surprising at all. Every single movement of the pandemic was a repetition of an old trope. When the original SARS emerged, it was Baric who broke down the epidemic into its three constituent parts, each defined by a set of key mutations: a zoonotic phase, when the virus separated itself from the many other bat viruses in circulation by infecting and replicating in a human host; an entrenchment phase, when mutations in progenitor virions allowed SARS, for the first time, to move from one human host to others; and an amplification phase, when mutations enhanced the virus's transmissibility, transforming it from a regional problem into a pathogen capable of causing epidemics across the globe.

As Baric observed the spread of SARS-CoV-2, he was struck by how closely the sequel hewed to the original. But while with SARS he had obtained genetic sequences of the virus only well into the epidemic, scientists had been sequencing and uploading SARS-CoV-2 genomes beginning in January 2020, with no sign of slowing down. Instead of working with a few dozen SARS sequences to reconstruct the pathogen's key mutations, the COVID-19 pandemic had, within a year, seen 119,982 different SARS-CoV-2 genomes publicly uploaded to the National Institutes of Health's GenBank. It was an astounding number, one that spoke both to the advance-

ments in genetic sequencing that had occurred since SARS first hit
and to how pervasive the scientific obsession with tracking the new
virus had become. Baric, of course, hadn't needed a pandemic to join
the ranks of the obsessed. Having spent four decades studying coronaviruses, he watched and waited for the patterns to repeat themselves, confident that this new virus would heed the same rules—of
evolution, of fitness, of the desire of all organisms, even those that
did not clear the threshold for life, to proliferate—and begin its antigenic drifts and shifts. The only thing that remained unknown was
whether the inevitable variants would be deadlier, more transmissible, or both.

The first variant of concern to arrive was the product of a random mutation in the original wild-type SARS-CoV-2 virus, which
transformed it into a more refined, more elegant, and more transmissible organic machine than its immediate ancestor. The variant
was dubbed D614G, shorthand for a seemingly innocuous swap of
molecules at the 614th amino acid on the spike protein located just
outside the virus's receptor-binding domain. At that point, a random replication error caused a bundle of sixteen atoms known as
D-amino acid to be removed in favor of the more compact ten-atom
G-amino acid. This minor atomic rearrangement, causing the virus
to net out with four fewer atoms, was nevertheless monumental. G
(the simplest amino acid, commonly found in fish and nuts and even
detected in a distant comet) made D614G's spike proteins more
turgid, causing the variant to have five times more intact spikes on
its surface compared to the original SARS-CoV-2 virus. More
spikes meant more chances to hook on to cell walls, which meant
more virions could force their way into cells. This subtle atomic rearrangement, it turned out, had global ramifications. As scientists
traced the genomic clues back, they realized that the emergence of
D614G had coincided with a sudden eruption of COVID-19 cases
across the globe at a time when the nascent pandemic looked as if
it may have been under control. As cases began to skyrocket across
the world, it was this variant that propelled them. By June 2020,
98 percent of cases in the United States were D614G variants, up

from 5 percent in the previous January, a story mirrored in every hemisphere of the globe. D614G had won the arms race.

Baric, perennially seeking to add to his viral menagerie, was intrigued. Intimately familiar with the thirty thousand–odd kilobases that make up coronavirus genomes, and an old hand at synthetic recombination, he did what came naturally. He engineered his own bespoke version of the variant, taking the original "wild-type" SARS-CoV-2 and swapping out the single consequential D-amino acid in his lab. Then he got some hamsters.

Baric's team placed thirty-two hamsters, identical in every regard, in individual cages, and then set them in pairs, with each pair just a few inches apart. They then injected eight of the hamsters with the ancestral virus and another eight with the spikier D614G variant. Then they waited. After two days, five of the eight hamsters paired to the D614G-infected ones had become infected. But the hamsters set a few inches from those infected with the ancestral SARS-CoV-2 stayed virus-free. Thankfully, when Baric's team followed the course of infection, they found that D614G didn't make the animals sicker; nor did it ravage human cell lines any more efficiently than the ancestral virus. D614G may have made the virus more transmissible, they concluded, but it wasn't making it any deadlier. Baric had revealed the evolving fitness of the virus captured in the breath of caged hamsters, the epidemic playing out within the confines of his BSL-3 lab. Though more efficient than its ancestor, the D614G variant was thankfully still susceptible to mask wearing, physical distancing, and lockdowns, the basic protocols that kept infections at bay. But it was also just the start.

July 2020, San Diego, California

Davey Smith was also trying to adapt to the pressure. Since the funding came down for his hydroxychloroquine clinical trial, he had been working flat-out to get the pieces in place. "I've never worked so hard in my life," he said, "not even when I was an intern, and I was young when I was an intern." His days typically started with a conference call at 5 A.M. with the National Institutes of Health to talk about the trial, with his last call of the day ending sometime after 8:30 P.M. Beyond juggling one of the most rapidly launched clinical trials ever, Smith was also an infectious disease physician, which meant he was also caring for the onslaught of COVID-19 patients admitted to UC San Diego's Jacobs Medical Center, a curvilinear oblong structure set back near the edge of campus. The cases Smith cared for ran the gamut from simple coughs to severe fevers; from mild breathing problems to pneumonia so extreme that the lungs of patients wouldn't even stretch, leaving them helplessly gasping for breath. "That was probably the hardest part, because one of the worst feelings in the world is feeling like you can't breathe. I'd see the panic in their eyes when they'd come into the hospital." It was all too much. "Before, I was running on adrenaline," he said. "But I can feel it now."

The heightened anxiety of the pandemic, along with his first-hand recognition of the need for high-quality data on treatments, motivated Smith even more to get the hydroxychloroquine trial running smoothly. But since its launch on May 1, 2020, he had had to contend with the fact that COVID-19 hadn't warped just time lines but scientific reality itself. Clinical trials are predicated on equipoise, the notion that neither of the arms of a trial (the drug arm or the placebo arm) is known to be better than the other. Without equipoise, a clinical trial is useless: if you're sure that the drug is going to do better than the placebo, testing it will only delay getting it to the patients who need it.

By May 2020, hydroxychloroquine had become a case study in the pervasiveness of COVID-19 misinformation, forcing Smith's attempts at achieving equipoise to run up against a seemingly intractable barrier. Every person, Smith was reminded in brutal fashion, had an inalienable right to hold an opinion, no matter how misinformed. And after President Trump's "What do you have to lose?" press conference in March, everybody, it seemed, had an opinion about hydroxychloroquine.

On May 7, a day when almost 1.25 million Americans had been

diagnosed with COVID-19 and the national death toll topped 75,000, Smith enrolled his first participant in the trial. The goal was 2,000 COVID-19–positive participants, half of whom would get hydroxychloroquine and half of whom would get a placebo. But over the next few weeks, as he tried to fill out the numbers in a desperate attempt to stay ahead of the rising death count, he was confronted with just how intractable people's beliefs about the drug had become. The pandemic had caused the United States to reach new heights of factionalism, the schism between those who supported the president and those who loathed him becoming wider each day. Smith was no stranger to politics intersecting with his work: any scientist who had cut their teeth working on the AIDS pandemic was intimately familiar with the way political opinions could be used as a cudgel to stop scientific discoveries that could help unpopular people (like gay men, people who injected drugs, sex workers, and migrants) from dying. In the HIV pandemic, it was the intense advocacy on the part of people dying of AIDS (especially gay men) that applied sufficient pressure to get lifesaving medicines to those who needed them. With HIV, politics was essential to moving the science forward. This rule, Smith discovered, had been turned on its head with COVID-19.

After moving at breakneck speed to get the hydroxychloroquine trial approved, getting participants enrolled was like slamming on the brakes. When Smith dug into why, in the midst of a pandemic for which humanity had no cures, people were unwilling to test out a potential lifesaver, the answer always came back to one word: Trump. In Smith's telling, half the people who were interested in enrolling loved Trump, and when they were told that there was only a 50 percent chance they would be given hydroxychloroquine, they balked. The drug, they were told by the president, would save their lives; why would they enroll if they weren't going to get the promised cure? The other half hated Trump and figured that if the president was hawking a miracle cure, it was not only too good to be true but would also very likely kill them. (At a White House briefing held a week prior to the launch of the hydroxychloroquine trial,

Trump had mused that disinfectant injected into the body, which is a really effective way to die quickly, could be used to kill the virus.) If Trump said hydroxychloroquine would save their lives, it would almost certainly do the opposite; why would they enroll if there was a chance they'd die?

Meanwhile, Smith's attempts at achieving clinical equipoise were crumbling as outside events took on a life of their own. The U.S. Food and Drug Administration's emergency-use authorization (which was not an endorsement of the drug's effectiveness per se, but only permission for clinicians to use it) had led to millions of doses of hydroxychloroquine being distributed to hospitalized COVID-19 patients. This made running a rigorous clinical trial of the drug nearly impossible. "It was so much out in the community that people just used it whenever and whatever," Smith said, "and that just sort of killed any chance of being able to have a good study of the drug to figure out if it worked or didn't work." The issues were manifold: the ease of access to hydroxychloroquine meant that trial participants randomly selected for the placebo arm might still get the drug prescribed to them by a doctor out in the real world, which would contaminate the data, rendering it practically useless. And if clinicians were going to use hydroxychloroquine regardless of the evidence, why bother spending millions on a trial?

Smith, his patience tested, was in a no-win situation. He had found himself trapped in one of the winding tributaries of scientific research on COVID-19, his work largely dismissed before the results even came in. As if to hit home just how much of an afterthought his sprawling hydroxychloroquine clinical trial had become, none other than Anthony Fauci, the director of the National Institute of Allergy and Infectious Diseases (which was sponsoring Smith's trial), had made public remarks disparaging the drug's effects before Smith's data had even been released. On May 27, 2020, less than a month after the trial's launch, Fauci told a CNN reporter that "the scientific data is really quite evident now about the lack of efficacy," and he suggested that hydroxychloroquine was too dangerous to use. Smith was dumbstruck. "So, Tony Fauci thinks hydroxychloroquine doesn't work," he said, clearly exasperated, a day later. "Really? Do you know something we don't?" Fauci was likely referring to the lower-quality observational studies that had already been released; still, it was the last thing Smith needed. Fauci's CNN pull quote had, it seemed to him, effectively prejudged the outcome of the trial.

Out in the field, the cacophony of voices weighing in on the purported benefits and risks of hydroxychloroquine soon proved to be too much. The opinions of Trump, Fauci, and the mix of inconsistent data about the drug caused Smith's lofty ambitions for a rapid trial to come crashing down to earth. By July 7, two months to the day after Smith enrolled his first participant, he decided to pull the plug. He put it plainly in his report to his funders at the NIH: "The study was terminated early," he wrote, "due to slow enrollment and lack of community enthusiasm." This dry text obscured the truly abysmal level of uptake he encountered out in the field. Though he had set a goal of recruiting two thousand people, and despite getting recruitment sites up and running in Alabama; California; Washington, D.C.; Illinois; North Carolina; Ohio; Pennsylvania; Texas; and the state of Washington, after two months he had convinced only 16 people to enroll—a whopping 1,884 participants short of his goal.

It was a crushing disappointment at a time when the world needed hope. For Smith, it was just another in a long line that stretched throughout his career. Loath to offer optimistic takes on the course of the COVID-19 pandemic, he nevertheless conveyed his dismay about the system breakdowns while wearing a wide grin. He was evidently familiar with the feeling of oscillating between a fervent desire to save lives and an awareness that most of what he and others did would be for naught. But steeped in failure as he was, he knew that moments of abject ruin almost always turned out to be brimming with much greater complexity, creativity, and potential than they appeared at first blush. Not that Smith would concede that within each failure lay potential success; that would be far too straightforwardly optimistic. Still, he understood that failure was the most fertile substance upon which science bloomed. And he was not alone.

IT MAY SEEM ODD THAT SCIENCE WOULD BE BUILT ON FAILURE. AND yet, the structure of this system of knowledge mirrors that of the universe, with elements of shining brilliance surrounded by vast regions of dark matter that, while invisible to the observer, make up the bulk (85 percent, in fact) of its mass. So it is with science: we often see only those luminous moments of genius, those elegant solutions so bright they saturate the light, which render us unable to glimpse the great mass of turgid and sorry failures that is the more abundant substance one encounters during scientific exploration. Failure, for most, is the name of the game; it is what scientists sign up for. Case in point: In 1984, the then secretary of the U.S. Department of Health and Human Services, Margaret Heckler, stated with absolute conviction that a vaccine for HIV would be available within two years. By 2021, with hundreds of millions of dollars spent and tens of millions of lives lost, those efforts had yielded exactly zero vaccines.

Failure as a scientific motivating principle is not restricted to the creation of vaccines or therapeutic drugs, nor is it just a comforting turn of phrase scientists use. If an experiment is successful, there's no need to continue finding new ways to solve a scientific problem. Success, in that sense, is a dead end. Failure, though, opens up a world of scientific questions that go beyond the binary. If a drug trial failed, was it because of the drug or of some flaw in the experiment's design? Could the drug's poor performance have had something to do with the people it was tested on? Or was it all, maybe, just random chance, and a repeat experiment would show great success? All these questions naturally arise after failure, like mushrooms after a rainstorm. And though they may differ in tone and subject, they all lead their questioners to the same place: the realm of doubt. And if failure is the driving force that causes scientists to ask new questions, doubt is the substance that causes them to steadily improve the ways in which they answer them.

The planned trials for the Moderna and the Pfizer mRNA vaccine candidates, both of which were designed to be incredibly massive, were exemplary of the role doubt played in pushing scientists to better quantify the discoveries they made. The enrollment targets for Moderna's candidate, built on technology designed by Derrick Rossi and adapted by Tal Zaks and others, was upward of 30,000 participants (though they could be enrolled only after Baric finished his safety studies with mice). The Pfizer-BioNTech trial, which used technology designed by Katalin Karikó and Drew Weissman, Rossi's twin inspirations, planned to enroll close to 45,000. Both were testing mRNA-based vaccines against COVID-19, and both vaccines had been seeded with Barney Graham's 2P spike protein, which held its shape and helped the immune system better match its antibodies to the coronavirus spike. With so many participants, these trials would provide strong evidence of, and thereby dispel doubts about, whether their respective vaccines actually worked against COVID-19.

Even so, Tal Zaks was convinced that one trial on its own wouldn't be enough to fully contain doubt about the viability of mRNA technology as a vaccine platform. "People ask me, 'What's the difference between your vaccine and Pfizer's vaccine?' And my answer to that is: 'That question is wrong.'" For Zaks, the best way to cast out scientific doubt was through repetition. "As scientists, we're taught that if you see one observation, it may be a fluke. Show it to me twice, three times, four times, and then I'm interested." Zaks pointed out that FDA guidance was that novel therapies and vaccines should have at least two or three Phase 3 trials. "And here we are, two large companies, both doing the same science in the sense that it's an mRNA, it encodes for the same spike protein, but we've no idea what goes into their secret sauce, and vice versa." Deeply invested in proving the promise of mRNA, Zaks wasn't interested in the differences between Moderna's and Pfizer's mRNA vaccines.

"The beauty is what was the same between us and Pfizer. That's the question." Taken in that light, Zaks had as much riding on the success of Pfizer's mRNA COVID-19 vaccine as he had on Moderna's. If both failed, the future of mRNA vaccines would be in great doubt. If one failed and the other succeeded, observers would surely doubt how replicable the platform might be. But if both succeeded . . . that was when doubt about the technology would disperse like dust.

Doubt is everywhere, always, unending. One of the least-understood gifts that the scientific method has given to the world is a way to wrap our arms around that doubt, to apply form to what is formless, to take our understanding of our ignorance and change that into a kind of clarity. The scientific method, almost miraculously, is a way to erect walls around doubt; to look through the fog of unknowing that envelops us and measure its density. Seen through that lens, science isn't just about discovery. It is the work of putting boundaries on our uncertainty. All that Baric, Lipkin, Graham, Smith, and others had done to advance humanity's understanding of coronaviruses and to develop cures was predicated on maximizing confidence in the validity of their discoveries. This didn't mean they sought to excise doubt completely—that would be a fool's errand. The trick instead was to minimize it enough that the results their experiments generated couldn't be explained by chance. And that was a task that scientists have been engaging in for a very long time.

Until as late as the nineteenth century, science was still just one among many competing systems of knowledge. While the scientific method had by then proven its ample value to society, having driven rapid advances in agricultural techniques, astronomy, magnetism, electricity, and many other disciplines, science shared a weakness with another major system of knowledge—namely, religion: the ideas it espoused had mostly to be taken on faith. The seeds of science's eventual dominance over religion were sown in the late seventeenth century by the development of statistical probability, which allowed science to take its exalted place in the center of the pantheon of knowledge.

While statistical probability sparked the eventual apotheosis of science, its originators weren't so high-minded: they were just interested in winning some money. The first book of statistical probability, published in 1657 and written in Latin by the Dutch scientist Christiaan Huygens, was called *Libellus de Ratiociniis in Ludo Aleae,* or *The Value of All Chances in Games of Fortune.* Huygens started the book off with a startling statement: "Although in games depending entirely upon Fortune, Success is always uncertain; yet it may be exactly determined at the same time, how much more likely one is to win than lose." From there, Huygens (one of the most consequential figures in the scientific revolution, who invented the pendulum clock and discovered Saturn's moons) went on to prove that games of chance were ruled by probability and to "explain particularly the Chances that belong more properly to Dice."

The book was a bombshell. No sooner had it been published than the *Libellus* found a wide audience desperate to understand how they could win more money at dice; if that meant learning a little probability theory along the way, so be it. Sales of Huygens's treatise on gambling then spurred the speedy adoption of probability theory among a growing number of disciplines (mathematics, actuarial science, and mechanics, among others) before becoming a central component of scientific experimentation.

Three hundred and fifty–odd years later, there's no going back. Just as Huygens was able to determine the rules of probability that governed repeated dice rolls, scientists nowadays apply probability to bridge the gap between the experiments they conduct in a lab and what the results of those tests might mean once they hit the real world. More than anything, the quantification of doubt that began with Huygens has transformed the binary distinction between scientific success and failure into a much blurrier landscape. And that's good news for us all.

In practical terms, every scientific study reports a point estimate, which is a single number that demonstrates the experiment's outcome, like a vaccine's level of efficacy in protecting against COVID-19. But studies also always present a range estimate, which is a range of values within which scientists are confident the true result falls. This might seem like a paltry distinction, but the difference between results expressed as a single number (the point estimate) and those expressed as a range of numbers (the range estimate) is as monumental a shift as thinking of the world in three dimensions or four. The two numbers that bookend the range estimate tell us all we need to know about the quality of the study and how much confidence, or doubt, we should have in its findings.

In the world of vaccine development, the range estimate is especially critical in weighing risks versus benefits. Case in point: If scientists reported that a vaccine for Norwalk virus, a debilitating and sometimes deadly gastrointestinal pathogen, could reduce infection by roughly 50 percent, that would be hailed as a breakthrough. (For context, 50 percent effectiveness is the cutoff the FDA has set to approve COVID-19 vaccines for distribution.) With a 50 percent point estimate in hand, government officials might be inclined to apply funds to purchase the vaccine in bulk, at the expense of funding other ways to prevent Norwalk virus, like prevention campaigns. If the range estimate of efficacy were also small—say, scientists reported that the vaccine's efficacy was between 42 percent and 53 percent—there might still be a strong case to make to purchase it. As the range estimate grows in size, though, the justification for investing in the vaccine becomes weaker and weaker. As a crude rule of thumb, range estimates get tighter the more people you enroll in a study; a vaccine with thirty thousand people will have a far narrower range estimate than one that enrolls only thirty. This makes intuitive sense: the closer we get to testing how a vaccine works across numbers comparable to the entire human population, the more confidence we will have in whether the results of the experimentation will mirror what happens in the real world. The flip side, of course, is that as the fraction of people being tested gets smaller, the range estimate gets larger, and confidence drops about whether the results are the product of a vaccine or of chance. And that's exactly the case with a real-life Norwalk vaccine candidate that was developed in 2011. Though its efficacy was 53 percent (above the 50 percent cutoff), the number of study participants was only ninety-eight, which had dire implications for the confidence the scientists had in their results. With the vaccine tested on such a small number of people, the range estimate extended from 16 percent to 74 percent, wide enough to cast serious doubt on whether it was conferring much protection at all. Sure, a vaccine that can halve your risk of developing Norwalk is useful, but one that reduces your chance of infection by only 16 percent? That's not much protection at all.

ON JUNE 15, 2020, WITH DAVEY SMITH'S TRIAL STILL TECHNICALLY running, the U.S. Food and Drug Administration revoked its emergency-use authorization for hydroxychloroquine, which it had provided in the weeks after President Trump first touted it as a game changer. Officially, the decision cited inconclusive data about the drug's benefits and evidence about its potential to cause heart attacks. Smith, had he had the will to fight on, would likely have taken issue with both these suggestions. But before the trial had even come to a close, he had already been dragged into his next all-consuming COVID-19 adventure, courtesy of his funders, the AIDS Clinical Trials Group. The ACTG, looking at a COVID-19 pandemic that showed no signs of abating and an almost complete lack of effective treatments for COVID-19, had hatched a plan, and Smith was to play a major role.

Despite the vast sums being spent on developing COVID-19 therapies, the list of treatments that had been authorized for use by June 2020 was very short: remdesivir. But that didn't mean that there weren't dozens, if not hundreds, of promising therapeutic drugs out there. The problem was that, with the pandemic death toll rising, they all needed to be tested as rapidly as possible to save lives. Running individual clinical trials for each potential COVID-19 therapy was a fool's errand. Instead, the ACTG decided, the only way to quickly amass an arsenal of effective treatments was to build a trial platform and then never shut it down. Under Smith's watchful gaze, the ACTG informed him, the ashes of the hydroxychloroquine trial were going to be transformed into something else entirely, with Smith in charge. Dubbed ACTIV-2, this adaptive clinical trials system would continually and rapidly cycle through the exhaustive list of therapies that might make a difference for people infected with COVID-19, with drugs being tested over a period of two weeks to two months before the next one was slotted in. It was, in some ways, the opposite approach to READDI: whereas that plan was meant to prepare humanity before the next pandemic spread out of control, ACTIV-2 was meant to generate solutions to the present crisis at the speed at which the crisis moved. Smith had gone from one failed drug trial to leading one that might not ever end. Despite his utter exhaustion, how could he say no?

This leap sent him into the oblique and chaotic world of Operation Warp Speed. A U.S. government program developed by the Trump administration at a cost of $18 billion, Operation Warp Speed was officially launched on May 15, 2020, as an audacious moon shot: an all-out effort to ignite the vaccine race, develop drugs that effectively treated people who were sick with COVID-19, and help the United States reclaim its place at the head of the world's technological superpowers. "If we can develop an atomic bomb in two and a half years and put a man on the moon in seven years," the head of Health and Human Services reportedly declared to his staff as the initiative was launched, "we can do this this year, in 2020." ACTIV-2 was a key part of the mission's activities. But as the numbers of the pandemic dead continued to mount throughout the summer of 2020, Smith saw the dark side of the moon shot: from his vantage point, Operation Warp Speed became a microcosm of market savagery.

With his elevation to principal investigator of the ACTIV-2 trial platform, Smith was suddenly thrust into the center of a swirling storm. Having worked at university hospitals and dealt with scientific peers at the National Institutes of Health throughout his entire career, he was shocked, as he logged into Operation Warp Speed teleconferences, by the kinds of people to whom he now found himself accountable. "When you're on calls with people whose first name starts with general or captain or admiral, that becomes a lot of pressure," he said. On his first call, his stomach dropped as he realized that eighty people were there to watch his presentation outlining the ACTIV-2 study protocol and his plan to analyze the reams of data it would generate about a long list of potential therapies. Worse, thirty of the people on the call were senior biostatisticians who had been brought in to scrutinize his numbers. "Thirty!" Smith cried. "I don't even know where you'd find thirty biostatisticians." But they were all there; all had reviewed the protocol beforehand, and they all had questions for him. For all the pressure, Smith saw the good. "Even though we were doing it fast, there was a lot of people double-checking and triple-checking our homework to make sure that it was going to work out."

The key to the adaptive trial's success was the efficiency of SARS-CoV-2. In the first weeks and months after it emerged, nobody—not even Ralph Baric—knew how tenacious the virus might be. As outbreaks became epidemics and then coalesced into a global pandemic, the stakes to find effective treatments were raised, but those rising case numbers also provided a vast pool of COVID-positive people available to enroll in the adaptive trial. This cold logic meant Smith spent his days tracking where case counts were highest and then working to secure clinical trial enrollment sites in those locations. As he had learned with the hydroxychloroquine study, leaving anything to chance could mean abject failure.

As ACTIV-2 moved forward, Smith grinding through multitudinous problems every day, he couldn't quite shake the feeling of being a small cog in the vast and complex political machinery of Operation Warp Speed. His timidity went so far that when he was sent an email from his government handlers requesting sign-off on his trial's massive $94 million budget, he instinctively ignored it. Those were sums he had never worked with before, and he figured that whoever was supposed to sign off on them was well above his pay grade. It was only days later, when the email was sent again, flagged as urgent, that he realized he was in charge. He could have been forgiven his confusion: he had requested only $14,000 as his salary for leading the trial, which he calculated worked out to about $3.50 per hour. It would fall far short of what he was owed in emotional hazard pay.

June 2020, University of North Carolina, Chapel Hill

The July 1 countdown to the clinical trial for Moderna's mRNA vaccine candidate had almost run its course. By then, Baric's laboratory had taken on the feel of a military field site, with a swelling personnel list, constant expansion into neighboring labs, and a scientific to-do list—all of it marked urgent—requiring around-the-clock experimentation. The lab was equipped with pressurized chambers, massive freezers to hold virus samples, and animal containment facilities, where mouse enclosures were lined up on shelves in neat rows. The ambient sound of fans venting air was a constant. Lab workers wore full-body Tyvek suits with portable air breathers to scrub the air, a thick apron, and two layers of latex gloves—Baric had learned the hard way that mice can bite through a single layer of glove but, even in cases when they draw blood, don't have the force to puncture two layers. As the world's preeminent coronavirus researcher, Baric had been pulled in a million directions as the pandemic had accelerated. He had taken on a harried look, his thin white hair flowing long at the back and lifting off the top of his head in wild tufts, his exhaustion obvious as he ran from lab bench to lab bench, monitoring progress and offering guidance wherever possible. Despite it all, he had the look of a man in his element, the adrenaline coursing through him as he transformed his great stores of knowledge into a brick-by-brick defense of humanity against the pathogen. One of the cornerstones was the task that Barney Graham had brought him: completing mouse trials of Moderna's mRNA-1273 vaccine candidate. Without those data, Graham explained, Phase 3 human trials couldn't launch as planned on July 1, 2020, which meant a viable vaccine might not be ready by the end of the calendar year, and more people would get infected, get sick, and die.

Baric ran with the assignment. Like so many of his experiments with vaccines and therapeutics against coronaviruses, this involved a stop-and-start rhythm. There was a flurry of early activity as he injected mice with a chimeric form of the SARS-CoV-2 virus designed to bind with mouse receptors instead of human ones, waited for an infection to take hold, and then administered the first dose of the vaccine. Then there was nothing to do but wait two weeks until the next dose was due, and then Baric and his team set the mice aside for another two weeks to let the vaccine's effects take hold. Once that had been done, it was time to collect samples. Baric's team took blood and removed the entire lungs from the mice—the blood in order to check for the virus and for the antibodies the vaccine should have produced, and the lungs to check for hemorrhage caused by the virus, which should be absent if the vaccine worked. Only then could the analysis be conducted to see if any differences existed between the mice that were and were not vaccinated and whether those results were due to chance or to the vaccine's actually being effective. All in all, the process took over three months, leaving Baric's team little room for error if they were going to make the deadline.

When Baric and his team finally analyzed the data, he found tantalizingly good news that Moderna's technology worked the way everyone had hoped it would. The mice vaccinated with mRNA-1273 were able to almost entirely clear the coronavirus from their lungs, giving it no chance to replicate and cause severe disease. When they stretched the experiment two more weeks (six weeks after the first dose) to see how long that immunity might last, they found that the vaccine's protection remained constant. It was cause for celebration. Even better, there was no sign that the vaccine was making the mice more ill. Baric, never one to rush things, nevertheless hit his mark, getting the data back to Graham in time to show that Moderna's vaccine was very likely to be not only safe but also effective.

With those data in hand, on July 1, 2020, the Phase 3 trial for mRNA-1273 was launched. And so, despite the chorus of voices that in the summer of 2020 decried mRNA-based vaccines as totally unproven, and that looked upon the Moderna trial as a failure in the making, Baric knew differently. "People say there's no data on these mRNA vaccines—that's just not true. There was a ton of data beforehand." He was the one who had collected it.

While the stock market valuations of the vaccine makers soared with every new announcement, and as journalists made deep dives into the revolutionary technologies the private sector was bringing to the search for a COVID cure, the contributions of Graham, McLellan, Baric, and others were hardly mentioned. In Graham's view, that was just fine. "It's not a problem for me," he said. "We became government scientists so that we could have access to resources to do the kind of work we like to do. It's a public servant type of job, and so if these products get out to help people, that is a big, big reward for us. And, you know," he added, "we feel like we have a little ownership over all the vaccines."

But in another sense, he conceded, he was worried that his lab and the others that did research within the confines of the National Institutes of Health (known as its "intramural" program) might be forgotten altogether. "One of the problems that has come up over these last twenty years at least is that Congress has gotten the idea that maybe the NIH's intramural program wasn't really useful anymore. They couldn't see any practical results coming out of intramural research, basic research; they kept asking, 'Why do we care about all this stuff?'" Graham figured that this skepticism resulted from the true story's not being told well enough. The fact remained that almost everything that was developed to fight COVID-19 and every other pathogen came from NIH-supported research.

Graham's answer to his own question—"Why do we care about all this stuff?"—was twofold: speed and rigor. Under Anthony Fauci's direction, the National Institute of Allergy and Infectious Diseases had in 2015 placed a bet on Moderna as a vaccine maker. In return, the company had entrusted the government scientists with running high-quality clinical trials and doing them fast. It was a bargain that, for Graham, was the best way to get vaccines into arms as quickly as possible. Baric's delivery of data showing that mRNA-1273 was safe and effective in mice had kept the plan on track. But even so, with confirmed cases of COVID-19 surpassing the ten-million mark on July 1, 2020, the day that the clinical trial for Moderna's vaccine candidate was launched, the finish line felt as distant as ever.

July 2020, Vancouver, British Columbia

The pandemic had forced many scientists to confront their own doubts and failures. But few felt they had let more opportunities slip away than Bob Brunham. His efforts to create a SARS vaccine powered by a publicly funded international consortium had been scuttled by bad timing and a self-diagnosed lack of tenacity. As SARS-CoV-2 overwhelmed the world, Brunham had found himself snowed under and outmaneuvered by the tens of thousands of new entrants hawking their solutions in the voracious COVID-19 science market. Despite forty years of vaccine making, including three promising SARS vaccine candidates fully tested on animals, Brunham had been cast aside. His last act of desperation, writing to Canadian prime minister Justin Trudeau's generic email account, seemed likely to generate nothing but false hope. And yet, miraculously, he received an answer.

Not by the Canadian prime minister; it was not that kind of miracle. Instead, Brunham was invited to meet with a senior bureaucrat in charge of pulling together Canada's nascent vaccine strategy. Brunham was overjoyed; it felt good to know that he still pulled some weight, and the phone call would be a shot to get his dream of a vaccine trial off the ground. When he joined the call, Brunham quickly made clear his concerns regarding the obsession with mRNA vaccines and the need to deploy trusted vaccine platforms to stop the pandemic. Having already created spike protein–based SARS vaccine candidates, he felt he could do it again with SARS-CoV-2. He just needed a shot.

The bureaucrat listened politely and asked some questions. But when Brunham pushed them on his vaccine proposal, they declined.

Brunham was deflated. Still, there was something, the bureaucrat said, that he could do for them: Would the renowned vaccinologist join Canada's Vaccine Task Force and decide which vaccines the country should buy to protect its 38 million citizens?

Brunham, who had moved through his career peripatetically, hadn't seen that coming. Being on a task force also wasn't how he thought he would be spending the pandemic. He wanted to be in the trenches generating the next cure. But as he thought about it, the idea became more appealing. Being on the task force would give him a front-row seat to the inner workings of every single vaccine candidate produced across the world. For someone driven by curiosity, who had seen the public sector's capacity to make vaccines wither away as his career progressed, immersing himself in the science and using it to build a plan for an entire country was a worthy consolation prize. How could he pass up the chance?

So, throughout the summer of 2020, Brunham joined ten other scientists to decide which vaccines Canada should spend its billions on. Having surveyed the field from afar, he was adamant that the mRNA vaccines wouldn't be winners at the finish line. As he took his place among the group, he readied himself to fight against mRNA vaccines in favor of the "classical" vaccine platforms that he was sure were the only viable path forward.

But a funny thing happened after Brunham logged on to the Canadian government's teleconference software and saw his square screen surrounded by those of the other task force members and the dozens of civil servants listening in. After brief introductory remarks, representatives from Moderna and Pfizer were invited on to the teleconference one at a time to make their pitches. Deeply skeptical, Brunham listened intently as the representatives ran through the data they had amassed on mRNA technologies, which went back over a decade. Brunham was then given access to aggregated findings so that he could see under the hood of the scientific data, mined over years, that the companies had amassed. His mind was utterly blown. Though the mRNA trials for COVID-19 hadn't yet been completed, he was shown the early safety data that Baric, whom he deeply admired, had generated in his mouse models for Moderna's mRNA-1273 candidate, along with the results from the multiple mRNA vaccines (including the MERS vaccine candidate) that Tal Zaks had overseen. Brunham was gobsmacked. "The mRNA vaccines," he declared, "may very well be the future of how we respond to pandemics." The technology was just deeply impressive.

Most people, having been forced to retreat from a position they've held fervently (and publicly) for years, would be forgiven for feeling stung. Not Brunham; he was delighted. "Data trumps any opinion I might have had," he said. "And that's the beauty of being on the task force: I got to look inside the cupboard, and I also got to ask really hard questions of the scientists." Within a few short months, his view of mRNA technology had gone from distraction to deliverance. Having burned with a desire to develop a vaccine, he was nevertheless happy with where he had landed: on the outside looking in, like Moses in the Promised Land, tasked with guiding his people to safe harbor. "In some sense, my entire career was preparing for this moment," he said. "I only wish I was a bit younger. But it's exciting."

By July, the task force had submitted its expert recommendations to the Canadian government. Soon the announcement came: Canada had purchased millions of doses of both Pfizer's and Moderna's mRNA vaccine. The decision, the country's procurement minister explained, had been guided chiefly by the collected opinions of the country's Vaccine Task Force. The task force's members, Brunham included, had advised that Canada commit to purchasing doses prior to the final results of the vaccine trials, a move that had become de rigueur during the pandemic as countries sought a future lifeline despite the uncertainty over whether mRNA vaccines worked. In a sign that the task force was all in on mRNA, the Canadian government was the first country in the world to pay cash, up front, for millions of doses of Moderna's mRNA-1273 vaccine candidate. Brunham's change of heart wasn't just a passing fancy. It was true love.

Chapter 13

YOU'RE SO NAÏVE

July 2020, Seattle, Washington

AS SHODDY CLINICAL CASE REPORTS AND POORLY DESIGNED STUD-ies proliferated, Nick Mark found himself at a crossroads. He was desperate for COVID-19 therapeutics backed by good science to save his patients' lives. And yet, most everywhere he looked, all he saw were hyperbolic claims about obscure and dubious cures. As a savvy clinician, he knew there were real weapons available hidden amid the silver bullets. It would just take some care to parse out the valid science from the bogus claims. And though the first months of the pandemic were marked by clinicians taking shots in the dark, by July 2020, that accumulation of shots (those that failed and those that hit their mark) had been aggregated into increasingly rigorous scientific evidence that was helping the front lines fight back. Among the most important developments was the understanding that COVID-19 was a biphasic disease, with respiratory symptoms early on as the virus replicated freely, followed by multi–organ system failure as the immune system launched a scorched-earth attack using cytokine bursts that targeted its own cells to try to rid the body of virus, and which sometimes led to death.

Mark pointed to the results from the RECOVERY (Randomised Evaluation of COVID-19 Therapy) study, a randomized trial of more than two thousand COVID-infected inpatients at hospitals in the United Kingdom, as critical in giving him the tools to deal with that second "scorched-earth" phase. The RECOVERY study had shown that patients on ventilators who were given the steroid dexamethasone were 35 percent less likely to die, though the drug made no difference for patients who weren't as sick. Because dexamethasone suppressed the immune response, it made sense that it would help very sick people whose cells were being destroyed by cytokine bursts. Mark had been operating under the working assumption that the steroid could potentially save lives, but he admitted to not being entirely sure. The RECOVERY trial had changed all that. "When the study findings came out, that switched our use of steroids from being a 'some-of-the-time intervention' to an 'all-the-time' one." It was still far from a cure, but it meant that fewer people were going to die. And that was the best that Mark could hope for: the crude fashioning of some blunt weapons to fight off the growing darkness while some light remained.

July 2020, University of California, San Diego

From his vantage point as a clinician-scientist in the age of COVID-19, Davey Smith had long hammered home the same refrain: the United States needed a COVID-19 testing strategy, and it needed one yesterday. The issue wasn't that quality tests hadn't been invented. Since 2003, when Ian Lipkin had to search for the ends of RNA fragments to create a rapid and sensitive test for SARS, technology had advanced sufficiently enough that tests for new viruses could be developed within just a few hours. No, the issue wasn't with the technology. It was, instead, the fact that companies that produced tests stood to make a fortune if they cornered the market on COVID-19 tests. And that introduced perverse incentives into efforts to test how far and wide the pandemic had spread across the United States.

"The three most important parts of this bill are testing, testing, testing," said Speaker of the House Nancy Pelosi on March 13,

2020, as she unveiled a congressional plan to make SARS-CoV-2 testing free for all Americans. "We can only defeat this outbreak if we have an accurate determination of its scale and scope, so we can pursue the appropriate, science-based response that is necessary," she continued. At the time, the best evidence of the power of testing came from South Korea. The country's quick response was chalked up to the country's MERS outbreak, which had infected almost two hundred people and killed thirty-six in 2015. After the outbreak, the South Korean government passed legislation allowing mass data scraping (from credit cards, mobile phone apps, and geotracking instruments) to reconstruct the movements of people found to be infected and to identify those with whom they had come into contact. The scars of the country's last coronavirus-induced trauma had, it seemed, prepared it well for the next one.

While much of the rest of the world was facing steep increases in cases, South Korea had managed to slow the daily case count to a few dozen; it had also done so without relying on the blunt instrument of mass lockdowns. Instead, the country of 50 million had blanketed its communities with coronavirus tests and then surgically engaged in isolating the infected, while quarantining only the network of potentially exposed people close to them. By March 17, 2020, South Korea had already carried out 270,000 tests—5,200 per million people. It was a massive public health effort that made the U.S. rollout—fewer than 200,000 tests for a country with a population six times the size of South Korea's—look paltry in comparison, and that was credited with keeping the South Korean case count among the lowest in the world.

In the United States, efforts to get a coordinated national testing strategy off the ground had, by July 2020, collapsed with a thud. Though the Trump administration had promised that tests would be rolled out in the millions by the end of March, it took until July 24, 2020, for daily tests to crest the one-million mark. While blame was widely ascribed to Trump's failure to lead, Davey Smith knew that there were other forces at play.

Smith had initially seen signs of hope that a national testing strategy could be transformed from political rhetoric into reality. At the beginning of the pandemic, the companies he dealt with were moving as fast as they could to help get tests out to whoever needed them. The companies also made sure that their tests were as standardized as possible, which was good news for Smith, who regularly used components from different suppliers at his hospital when some part or another couldn't arrive quickly enough. The vials, saline solution, nasal swabs, the polymerase chain reaction kits used to replicate viral sequences, the various liquids used to transport samples, and the costly instruments that analyzed the samples—at the start, it didn't matter if they were mixed and matched, which helped enormously as hospitals, clinics, and public health units tried desperately to get a handle on how far the pandemic had spread and who was spreading it. If swabs sent by Roche were being analyzed using instruments from Abbott Molecular, all the better. Testing, all parties seemed to agree, was a matter of public good.

By the summer, any sense of collective action had disappeared. "Testing is a fiasco," Smith declared with visible anger. At the beginning of the pandemic, companies that produced medical diagnostic equipment had allowed customers like Smith to share components across different brands. But one day, when Smith tried to analyze the results of a test kit from one company using a device provided by another, he got an error reading. The companies had, it appeared, quietly introduced proprietary features into all their testing materials so that they could be used only with their own devices. This refusal to cross-pollinate meant that Smith and clinicians like him were hamstrung in the case of shipping delays, of which there were many. He had for years been able to use a Roche swab with Abbott instruments, but right when the volume of viral tests he was conducting was greater than ever before, this didn't work anymore. What was most galling was that all the companies had developed the same kinds of COVID-19 tests, so it didn't matter which company's components you used. Still, the companies had evidently bet on a new strategy to gain a larger share of the lucrative COVID-19 testing market. "They put their special sauce in," Smith said. "Not to make it better, mind you; it was to make sure you could only use their platform. It's the printer strategy: if you buy my printer, you can only use my ink."

The upshot was that Smith was entirely dependent on shipments from one manufacturer to keep his testing regimen going. Though the companies publicly declared themselves ready and willing to ship tests to whoever needed them, in Smith's experience this wasn't true. As head of UC San Diego's Division of Infectious Diseases, he had been pleading with Abbott and Roche to help meet the university's daily testing goals. In response, sales reps from both companies assured Smith that they were ready to scale up testing to whatever level he needed. But when UC San Diego tried to buy fifteen thousand tests per day, Abbott had a quick response: no. The simple fact was that testing companies couldn't meet demand for COVID tests, but none of them were willing to admit this lest they cede the market to their competitors. And this had left the government, clinicians, and research labs bereft and flying in the dark as new outbreaks cropped up. It was infuriating.

The testing debacle was only one part of a system-wide failure. Everywhere Smith turned, the enthusiastic selflessness the pandemic had early on generated in scientists was by the summer months transformed into a crueler version of for-profit science making. Evidently, old habits died hard, even when an existential enemy could be vanquished only through collective action.

Much to Smith's surprise, ruthless individualism had even pervaded Operation Warp Speed and wormed its way into the adaptive trial he was leading. ACTIV-2, set up to do rolling tests of therapeutics for outpatients infected with COVID-19, was in many ways a gift to private companies. As a public-private partnership, it was entirely funded by the U.S. government, but the companies involved were able to keep all the intellectual property attached to the drugs that Smith was testing for them. All they had to do was supply drug candidates, and Smith and his team did the rest, distributing the therapies to participants across the roughly one hundred clinical trial sites they had set up around the world. They then shared the data with the companies, who could market the successful products to the world. There was no better deal out there. The bottom line was that the companies were all going to be able to make money without risking anything. The only caveat was that they had to share some basic information with the others involved in the study (Smith's team, the military brass running Operation Warp Speed, and the other companies involved) so that the trial could run smoothly. It seemed like a perfect plan, the kind that would go off the rails only through self-sabotage.

But then, to his surprise, ACTIV-2 did go off the rails. At the end of a routine teleconference call a few weeks before the launch of the trial, Smith asked a straightforward question about how one of the antibody therapies he would be testing was presumed to work. Where, he asked, did the antibodies bind with the virus? It was the kind of question that, as the trial lead, he would obviously need to know. He was met with awkward silence. Eventually, a rep for the company that had developed the product offered a nonanswer. "I can't tell you where the antibodies bind to the virus," he said tersely, as the rest of the teleconference members, numbering in the dozens and including many of the company's competitors, listened in. Smith rolled his eyes. "You can't tell me or you *can't* tell me?" he responded. As calmly as he could, Smith explained to the company rep that knowing which part of the SARS-CoV-2 virus the antibody was supposed to target was nonnegotiable. "I need to know where to look for areas of drug resistance," he explained; that way they could stop the study before the virus evolved to outsmart the drug, which could have disastrous consequences. The rep steadfastly refused, but Smith would not budge. Eventually, the company's CEO came on the call and agreed to talk.

It was classic protectionist behavior by companies that had only ever known one way of doing business. They were "pocketbook defense mechanisms," Smith said with disdain, referring to the small-minded thinking that had no place in the grand saga of the pandemic. "'Oh, Davey, you're so naïve'" was the response he got when he confronted one of his colleagues about the episode. The label fit, Smith admitted, but who could blame him? As a scientist firmly entrenched in academia, he hadn't ever been close enough to pharma companies to know the bruising rules of the game. That had all changed during COVID-19.

It got worse. The petty protectionism could not prepare Smith for the bomb throwing that started after the trial was launched in earnest. ACTIV-2 was predicated on a fair deal and zero risk for everyone involved. Though the first organizations to sign up would be the first to have their drugs tested, all would eventually have the full resources of the U.S. government evaluating their medicines. Sure, companies farther down the queue would have to wait a few months, but each drug was moved through the adaptive trial at record speed; Smith himself marveled as one drug sailed through a Phase 2 trial (which tested the safety, side effects, and optimal doses of treatment) within just one week. This made the short waiting period that companies had to put up with before their drugs were tested well worth it. After all, there was more than enough room in the massive COVID-19 market for everyone's products. More to the point, the trial's purpose was to find cures, not make any one company rich. As far as Smith was concerned, in this everyone agreed. But that wasn't exactly true.

By June 2020, through sheer force of will, Smith had managed to get thirty-one sites across the United States up and running to test hydroxychloroquine. But when the trial switched from a single drug to a potentially unending series of therapies, his masters at Operation Warp Speed told him that he would have to increase the number of sites to at least ninety-five. It made sense. Smith had, after all, been able to enroll only sixteen people into the hydroxychloroquine trial, thousands short of his goal. He would need even more for the adaptive trial, because every drug he tested would require up to two thousand people: a thousand who would receive it and another thousand who would get a placebo. The more drugs in the trial's pipeline, the more people Smith would have to enroll. And while he had never been convinced by the talk of hydroxychloroquine as a game changer, the slew of medicines in the queue for ACTIV-2 just might be.

The first class of drugs Smith was testing was monoclonal antibodies. These were synthetically created immune cell replicas that flooded the immune system with highly targeted protection, like mercenaries brought in when the standing army was outmatched. While monoclonal antibodies were promising treatments, they were also prone to generating resistance among the viruses they attacked, and for that reason, they generally had a short shelf life. The viral proofreader made coronaviruses mutate at a slower pace than other RNA viruses like influenza, which meant that SARS-CoV-2 might not be able to rapidly generate resistance to monoclonal antibodies. But nobody knew for sure. Despite their potential drawbacks, Smith had been tasked with testing three: one from the pharma giant Eli Lilly, one from the Chinese biotech firm Brii Biosciences, and one from the United Kingdom–based pharma giant AstraZeneca.

Then there was an inhaled interferon drug, interferons making up a class of proteins naturally produced by the body that interfere with viral replication and summon the immune system to boost its defenses; interferons are also responsible for the fevers, headaches, muscle pain, and other flulike symptoms you feel when a virus has invaded your body. Interferons belonged to the larger group of proteins known as cytokines, which the immune system bombarded the body with in the second, scorched-earth phase of COVID-19 disease. While the release of cytokines sometimes ended up killing COVID-infected people, the theory was that in smaller doses, they would still destroy the virus without harming patients.

Next on the list was a pill-based protease inhibitor, which belonged to a class of antiviral drugs developed primarily to treat HIV infection. Proteases are enzymes that break down large macromolecules into smaller components, and they're employed by viruses to fashion the pieces used to build copies of themselves. A protease inhibitor, the theory went, might work against SARS-CoV-2 by targeting its capacity to assemble new virions.

The stakes, for Smith and for the world, were high. Any of the medicines, and potentially all of them, might be a cure. But those drugs were just the beginning, and the list of interested companies and potential cures just kept getting longer. It was enough to keep Smith busy for years. To keep up with the demand, he had to get his trial sites up and running fast. It was, of course, the companies themselves that stood to benefit the most from getting the sites online. The more people who enrolled, the faster the companies could prove that their drugs worked and the faster they would all move to market. As Smith knew, though, finding trial sites wasn't straightforward, because the team running each site needed a highly specialized set of skills to run clinical trials. The sites also needed to be reasonably far apart, ideally in different cities, so that they wouldn't inadvertently scoop up each other's participants. This left Smith with little wiggle room, and he went so far as seeking out what he called "guns for hire": private clinics that specialized in running for-profit trials, idly waiting for the next contract to come along.

Smith had banked on collective action to keep the competing pharma and biotech companies involved in his study committed to its success. After all, the trial's speed and flexibility benefited all of them as they sought to get their drugs to market. But one day, on routine calls with his local trial sites, he got word that some of the sites had received offers to switch their allegiance from ACTIV-2 to competing drug trials. This was galling in its own right, but there was more to the story: the companies that were trying to poach the sites for other trials were the ones that had already signed up to have ACTIV-2 test their drugs. Smith was furious: members of his own team were trying to undermine his study's success, which struck him as absolutely absurd and just plain vindictive. Why would pharma companies, who benefited the most from having Smith's trial up and running, be trying to slow it down by peeling off places where participants could enroll and test their drugs?

As Smith consulted the more seasoned trialists on the team, they again expressed their surprise at his naïveté. The internal poaching was totally predictable: though they were all ostensibly working together, pharma companies would never stop acting in their own self-interest. Companies whose medicines were being tested a few months down the line were highly motivated to throw sand in the works to slow down their competitors whose drugs were being tested before theirs—even if that meant that they themselves would suffer later. "Oh, it happens all the time," Smith was told. Shockingly, it wasn't just one pharma company among the group, and it wasn't just the for-profit ones. Smith discovered that even the Bill and Melinda Gates Foundation, which was a key partner of the AIDS Clinical Trial Group, which ran ACTIV-2, and was also the largest nonprofit supporter of research on vaccines and therapeutics, had been poaching sites for other trials it was involved in. "It's all cutthroat," he said, clearly incensed. Making matters worse, companies that had flirted with joining the ACTIV-2 trial but that had ultimately decided not to were also picking off trial sites and holding them so that their competitors in the trial couldn't use them.

It was all too much. Smith's trial platform simplified every stage of the process for testing drugs and did so at record speed. He had set up sites all across the country and around the world. He fervently believed that if he could keep the enterprise going, pharma companies would get their drugs approved quickly, which meant patients without treatments for COVID-19 would finally, if belatedly, get access to cures. On top of that, the companies would get rich. And yet, for all his platform's benefits, he seriously began to doubt its chances. In an ideal world, Smith could test various drugs for the next two years to try to keep people out of the hospital. "So, we can take any drug, put it into the trial, and have an answer relatively quickly. That's great." Still, this obvious collective good, he worried, wouldn't be enough to save ACTIV-2. "I don't know if it's going to continue to work because, for lack of a better term, it requires a bit of socialism," he said. "It's the only way to keep the trial in place." Once the initial anxiety over the pandemic was over, he predicted, scientific research would revert completely to its capitalist roots.

One company, Merck, had jumped ship from the ACTIV-2 trial after just a few weeks. "They were very interested," Smith said, "and then they pulled out. Now they're struggling to sort of get it together." The talks with Merck were ongoing, but Smith wasn't optimistic. The company seemed impatient to test its miracle drug, just like everybody else. It was a nucleoside analog that acted in the same way as remdesivir, by swapping itself into the SARS-CoV-2 genome during replication, camouflaging itself from the proofreader, and creating a cascade of errors that ultimately caused the nascent viral architecture to collapse in on itself. The best part? EIDD-2801 could be taken as a pill.

Smith, having seen the drug's profile, was keen on testing EIDD-2801 in ACTIV-2, but he didn't have the time or energy to harangue pharma companies who didn't know whether they were coming or going. Ultimately, he was resigned to Merck's going its own way. "I think it might be a good drug, but who knows? We'll see what happens when they test it. Ralph Baric was involved," Smith said in passing; it was clear that Baric's involvement made the drug worth pursuing. By late October 2020, four months after Baric and his collaborators had quietly launched a clinical trial of roughly two hundred people to test EIDD-2801, Merck finally launched its own in-house trial with a target enrollment of 1,850 participants.

Though Smith had relished the chance at testing Baric's brainchild, Merck's decision to go it alone meant that, like the rest of the world, he would just have to wait and see whether it would live up to its hype.

Chapter 14

THE QUESTION IS GOING TO
BE THE FRINGES

October 1, 2020, San Diego

IT WAS, DAVEY SMITH ADMITTED, A NOVELTY. THERE HE WAS, HAV-
ing dinner at a restaurant, laughing, surrounded by people he
loved, celebrating a friend's birthday. It was the hottest day in weeks,
a scorching 97 degrees Fahrenheit, but this was San Diego, land of
perennial warmth, a desert dressed up like a verdant theme park, a
city where seasons barely existed. In any case, 2020 was the year
humanity stopped marking the passage of time with changes in the
weather and looked, instead, to seven-day trends in new COVID
cases and fatality rates to track the passage of time. And though it
was the fall, those trends bore the hallmarks of spring: hope.

The positivity rate in San Diego County had dipped to only
3 percent, suggesting a slow-moving pandemic unlikely to surge
anytime soon. Meanwhile, there were only two deaths that day,
proof that Smith and his colleagues had also begun to master the
care of COVID-19 patients after a long few months of simply trying to understand the course of illness. This good news had allowed
the hospital where he worked, the Jacobs Medical Center at UC San
Diego, to gradually return to a semblance of normality. The surge of
COVID-19 patients from the summer was finally dwindling, and
Smith had been given a much-needed break from his clinical duties.

Though he was still managing ACTIV-2, the adaptive clinical trial running across almost a hundred sites, even that had been rolling along nicely after the backstabbing between companies had simmered down.

By October 2020, he was testing bamlanivimab, or "bam" for short, a prized monoclonal antibody developed by John Bamforth's old employer, the pharma giant Eli Lilly. A synthetic antibody, bam was, like all monoclonal antibodies, originally harvested from a recovered COVID-19 patient, then purified and replicated in a lab. When injected into an infected patient, the monoclonal antibodies acted like a robot army primed for battle against the virus, reinforcements for a person's natural antibody response. Half the participants enrolled in the trial would receive an injection of bam; the other half would receive a placebo.

While antibodies work in many different ways to attack pathogens, bam specifically attached itself to a small section of the SARS-CoV-2 spike protein, which then rendered the virus unable to bind to cell walls. Monoclonal antibodies were a promising technology, especially for dealing with RNA viruses. In 2019, a new monoclonal antibody dubbed UB-421 to treat HIV was shown to be just as effective and longer-lasting than a cocktail of antiretroviral drugs, not to mention less toxic, which opened the door for a whole new class of treatments for the tenacious retrovirus. Bam, Eli Lilly hoped, could do something similar with SARS-CoV-2. And while there had been the predictable media stories about bam as yet another COVID miracle cure, the trial was still in progress, leaving Eli Lilly and the rest of the world holding their breath. Smith, knowing that the bam trial was under way without a hitch, was breathing easy for the first time in months. That night, as he gazed upon the faces of his friends across the restaurant table, thankful that the pandemic had spared them, he let himself relax into the moment.

And that's when he got the call. He looked at his phone: it was the U.S. National Institutes of Health. The caller, a government employee, wasted no time. "Somebody at the White House wants the drug."

Smith sighed. Earlier in the day, he had seen the news reports that Hope Hicks, a trusted advisor to President Trump and his administration's former communications director, had tested positive for the virus. The prior week had been a whirlwind of travel and events for Hicks, the president, and his administration. It started the previous Saturday, September 26, with a 150-person ceremony in the White House's Rose Garden to announce Trump's Supreme Court nominee, Amy Coney Barrett, during which masks were not worn and guests mingled in close quarters and after which eleven attendees, including multiple Republican senators and White House staff, tested positive for COVID-19. (Hicks, though, was not present.) Later that day, Hicks traveled with the president to an outdoor rally in Middleton, Pennsylvania, which was attended by more than two thousand people. Two days later, Hicks boarded Air Force One with Trump to Cleveland, where she attended the presidential debate. The following day, September 30, she was again on Air Force One, this time traveling to Minneapolis for a presidential fundraiser. The next day, October 1, was Davey Smith's friend's birthday.

As principal investigator of ACTIV-2, Smith was the only person in the world with the authority to allow someone not enrolled through one of the trial sites to use the drug, as it hadn't yet been approved by the FDA. This wasn't, of course, the first time the administration had intruded on his research. It had been only five months since Trump's touting of hydroxychloroquine as a COVID-19 "game changer" effectively ended Smith's trial of the antimalarial drug before it even got off the ground. He could have been forgiven, then, for feeling unmoved by the request.

Smith responded as he figured any scientist would: by protecting the integrity of his study at all costs. "Well," he said, "she can join my trial." He was well aware that this might come off as hubris, but he didn't care. Ensuring that the trial was done right was more important than letting the treatment be used off-label by one person, no matter how well-connected they might be. If Hicks enrolled in the trial, she would also have a 50 percent chance of getting bam, which

Smith thought was totally reasonable. Having a high-ranking administration official enroll in a clinical trial could also give the science a nice public relations boost. The government employee disagreed and began to argue with him, but Smith cut him off. "No," he said, "we don't give the drug outside of a trial." He pointed out that bam wasn't approved even for compassionate use because there was no data showing that it actually worked. Unlike the hydroxychloroquine trial, which had collapsed in on itself in the wake of media stories decrying the drug as either a death sentence or a miracle cure, Smith had maintained the bam trial's clinical equipoise—its effectiveness wasn't biased one way or the other—and he was intent on keeping it that way. So, as politely as he could, he declined, hung up, and returned to the birthday party.

Half an hour later, he was walking home in the desert night's waning heat, passing familiar palm trees and succulent gardens, when his phone rang. It was the National Institutes of Health again, but this time it wasn't just about Hope Hicks. "There's a few people we need to find the drug for now," he was told. Frustrated by the political pressure and burned out after months of navigating endless bullshit, Smith had had enough. It didn't matter how many people wanted bam: he wasn't going to jeopardize his trial by handing it out to anyone who wasn't enrolled. So, exploring the boundaries of his patience and politeness, he again suggested that whoever wanted bam should sign up for the trial. He even gave the address of the trial site closest to the White House: the Walter Reed Army Institute of Research, which was just a thirty-minute drive away. The phone call ended abruptly. "Oh well," Smith thought. It seemed that some people just couldn't be convinced to participate in cutting-edge scientific research.

When he woke up the next morning, he was greeted by bewildering news. President Donald Trump had, in the early hours of the morning, revealed that he had tested positive for COVID-19. Later that day, the president was admitted to Walter Reed National Military Medical Center, where he would ultimately remain as an inpatient for four days. The late-night call wasn't, Smith immediately realized, about Hope Hicks after all; it was about the health of the president of the United States.

Though Trump's condition and treatment were initially shrouded in secrecy, details soon emerged: he had been provided supplemental oxygen and had been given a barrage of experimental treatments. Upon his return to the White House the following Wednesday, President Trump gave an impromptu TV interview in which he singled out praise for the key treatment he figured had saved his life: a monoclonal antibody cocktail being tested in a clinical trial by the biotech company Regeneron Pharmaceuticals. Smith couldn't help but laugh. After getting denied access to bam, the White House had evidently sought out the next most-promising monoclonal therapy; unlike Smith, Regeneron appeared happy to compromise its clinical trial for some high-profile press. "Some people don't know how to define therapeutic," Trump claimed in the interview, in a tacit acknowledgment of his own ignorance on the subject. "I view it different: it's a cure." The president then went on to promise that, just as he had with hydroxychloroquine, anybody who was hospitalized with COVID-19 would be able to access the drug he'd been given. "I want to get for you what I got. And I'm going to make it free; you're not going to have to pay for it."

To Smith's dismay, the president then, in a seemingly off-the-cuff remark, obliquely mentioned bam as well. It, too, was a cure, he said, and would be widely and freely available to all who needed it. Smith was well aware of the damage that could be wrought by a president putting his thumb on the scale. He also knew it could have been far worse.

What the president didn't know was that Regeneron's monoclonal antibody cocktail probably didn't help him at all and more than likely hindered his recovery. On October 30, three weeks after Trump's speech praising the Regeneron product, the clinical trial testing its benefits was halted because of safety concerns. All clinical trials have an independent data-monitoring committee, an arm's-length group made up of scientists who are the only ones with unfettered access to the trial's findings; the committee is there to monitor adverse events and halt a trial if it becomes too dangerous. The committee assigned to the Regeneron trial had seen a disturbing signal among a subset of the 524 participants enrolled in the trial who were hospitalized with COVID-19 and who needed to be ventilated or required "high-flow oxygen" (essentially, forced air delivered to the patient). When they were given Regeneron's antibody cocktail, the drug appeared to make them sicker. In the clinical trial world, this was known as a "safety signal"—that is, a trend in the data linking the drug under study to disturbing outcomes among its recipients. "And that would have been the president," Smith said, "because he was sick enough to have to go into the hospital and get oxygen."

When the news came, Smith felt vindicated for his decision to deny the world's most powerful man an experimental drug when its benefits and harms were still unknown. "It would have been safer for him to go into the clinical trial," Smith said, where he would have only had a 50 percent chance of getting the Regeneron antibodies. "And he could have gone out to the world and said, 'Hey, I'm trying to help all of us by participating in the clinical trial.' But he never did."

By November 2020, new cases of COVID-19 had been surging relentlessly across the globe, reaching 4 million per week. It was an unfathomable number, made worse by the accompanying spike in COVID-19 mortality: between the beginning of October and mid-November, the weekly number of deaths across the globe had doubled, from 37,000 to over 65,000. The cruel irony was that the case fatality rate (the proportion of people dying of COVID-19) had actually been declining since April 2020. Back then, while the epidemic was still a largely unknown entity, 7 percent of people who were infected were dying. By November, as clinicians like Nick Mark improved their care, and as testing expanded to capture a larger swath of the COVID-positive population, the case fatality rate had dropped to 2.85 percent. Nevertheless, the overall number of dead kept rising. Meanwhile, Smith's ACTIV-2 trial hadn't yet identified any drugs that would revolutionize care for people who had become sick with the virus. The sad fact was that the spread of infection was far outpacing the improvements in care, like a loping hare outracing an exhausted fox as the bitter cold of winter set in.

It was an old story. Since the dawn of humanity—since before, since the first cells appeared on earth and perhaps even before then—the fittest emergent viruses have decimated human hosts, leaving us no recourse except to hold on and wait, under some form of quarantine or lockdown, until they burned their way through the population. In the case of the Northern Song, it took at least eighty years for a conflagration of epidemics to subside. The Russian flu, that probable coronavirus epidemic, killed one million people across the planet in roughly one year. The 1918 Spanish flu infected a third of the planet and culled one hundred million people before it faded two years later, spent, leaving in its wake an immune and traumatized species. The list goes on. Every single novel pathogen, upon its emergence, had either successfully dominated our species or been undone by its own lack of fitness—until SARS-CoV-2.

To those mired in it, the pandemic felt new, a brusque reconstitution of the social order along lines expressly designed to yield maximal anxiety and pain. The pandemic's daily Sisyphean task played itself out the same way for many: as a struggle to overcome the desire to be close to the ones you loved lest they inadvertently kill you, or you them, with the virus. Across the world, law and order devolved into smaller and smaller constituent parts: First, the borders between countries were closed and restrictions on movement across cities were enacted. Then, shelter-in-place orders came into effect, tightening even further the loop of movement that pandemic-era society allowed. It was life accompanied by dirge, the numbers of the sick and dead posted daily to remind everyone of their collective responsibility and guilt, the predictability of the rising deaths numbing the collective psyche, hollowing out the resolve to stay at home and creating a new class of amorality, as people flouted the rules of public health and safety, thirsting for the furtive touch of another person. It felt wholly, horribly novel. And yet, COVID-19 resembled every single pandemic that had come before. The virus weaponized humanity's need for intimacy, and by doing so, it transformed love into a deadly, selfish act. It punished those who did not have private homes within which to isolate themselves. And it sparked a furtive paranoia about the "true" origins of the virus, as if the collision between a novel pathogen and humanity were an aberration rather than, as was evident across the timescale of epidemic history, an inevitable coming together. As Ian Lipkin would have said: none of this was new.

What was new—what was fantastically, mercifully different— was the lowering of a scientific deus ex machina before the epidemic had finished its initial sprint through our species. In the midst of the global proliferation of the virus and its variant descendants, into the third act of the morbid theater of death and illness into which the world had been plunged since March 2020, interim data from the vaccine trials were finally delivered. By November 2020, a mere nine months after a global public health emergency was declared by the World Health Organization, against all odds, humanity had finally equipped itself with a shield. It was a remarkable feat, and though the vaccines seemed to have appeared almost spontaneously, their delivery into the world was a triumph decades in the making, as Barney Graham, Tal Zaks, Ralph Baric, and the many other virologists, vaccinologists, and molecular biologists working on them would attest.

Like so many scientific questions, it was a matter of degree. Some "classical" vaccines, like the influenza vaccine, were only on average 67 percent effective, though this figure could dip to as low as 20 percent in some years depending on the match between the vaccine's antigen and that season's particular flu strains. And yet, despite this low overall efficacy, and its wild swings from year to year, the flu vaccine nevertheless managed to prevent millions of flu cases each year. This proved such a winning strategy that the 2020 flu shot, which was only 45 percent effective, was touted as a vast improvement over those of previous years and was estimated to have saved hundreds of thousands of lives. Taking a cue from the flu, since the first whispers that it might just be possible to bring multiple COVID-19 vaccines to market amid the pandemic, the U.S. Food and Drug Administration had repeated the same refrain: the agency would approve only those vaccines that demonstrated at least 50 percent efficacy. This threshold might appear low, but what it revealed was the oft-ignored core truth about vaccines: they aren't meant to protect a person from disease.

If getting an appendix removed saved only half the people with appendicitis, the surgery would be discontinued. If there were only a fifty-fifty chance that getting glasses improved your vision, no one would shell out hundreds of dollars for lenses and frames. And yet, people regularly get vaccinated under the assumption that doing so will directly help them. As we know with the flu vaccine, that's not always the case. But vaccines aren't meant to protect a person from disease—what they are meant to do is protect populations from epidemics. Though that might sound like semantic quibbling, it speaks to the ways in which effectiveness is a relative term, wholly dependent on the magnitude of the playing field.

Vaccines are like getting eyeglasses if your eyesight magically improved every time someone else bought a pair. So it is with vaccines: a flu shot that's only 45 percent effective is a pretty useless medical treatment. But if everyone receives a shot, then 45 percent of the population is no longer able to spread the flu virus. If another quarter of the population developed natural immunity from having previously been infected, then that would exceed the rough threshold needed for herd immunity to the flu (roughly 65 percent of a susceptible population), at which point enough people would be unable to transmit the virus that it would stop circulating widely enough to reach you and make you sick.

Within the FDA threshold lay a message of hope. COVID-19 vaccines that exceeded 50 percent efficacy would give society more breathing room to contain the pandemic, as the number of people requiring vaccination would go down. A 70 percent effective vaccine would require only about three-quarters of the population to be vaccinated. With a 95 percent effective vaccine, only as few as 60 percent of people might need to get a shot. But if the vaccines failed to clear 50 percent? They were more potato guns than silver bullets.

Vaccine makers both in the private sector and at the National Institutes of Health had become bullish on the time line as the trials got up and running. This was largely because the trials needed at least some of their participants to get sick with COVID-19 to test how much case counts differed among those in the vaccine and placebo arms. With the pandemic's acceleration in the United States and elsewhere, new cases were not a problem. The potential efficacy of the vaccines, though, was less of a sure thing. But when the data were finally released by the independent monitoring committees overseeing the trials, those who had helped lead them (Graham, Zaks, and Baric) were floored. In the second week of November, results from the trial of Moderna's mRNA-1273 vaccine candidate, which Graham had overseen and which had enrolled more than thirty thousand people, were finally released. Moderna's vaccine, to Graham and Zaks's immense joy and relief, was close to 95 percent efficacious, which meant that reaching the herd immunity threshold of 70 percent of people either immunized or recovered from infection with SARS-CoV-2 was entirely possible.

Tal Zaks, Moderna's chief medical officer, was thrilled, though his long-standing confidence in Moderna's technology had made the results less of a surprise. Where Zaks became rapturous was when he compared the results of Moderna's trial with those of the Pfizer trial. Obsessed with proving that mRNA was a viable vaccine platform, he saw the hyped-up competition between the two vaccine makers and their products as beyond irrelevant. "Here we are: they had forty-four thousand participants, and we had thirty thousand," he said. "The results came within a week of each other, and both trials used a similar end point," which was preventing COVID-related illness, "and the efficacy results are within one percent of each other." Those almost identical results were the credibility that Zaks had so long desired. He wanted Moderna's COVID vaccine to succeed, of course. But beyond that, he wanted the public to understand just how much promise lay in the mRNA technology that underpinned it—which could be done only by showing that the effectiveness of Moderna's vaccine wasn't a one-off. "That's the most astounding scientific replication I've seen in my career, and the most important one. And it clearly demarcates mRNA technologies from all the rest of the pack."

While Zaks considered the Moderna and Pfizer vaccines as two similar experiments that produced identical results, there were differences that went beyond the vaccines themselves. Pfizer's was a wholly private effort, with the company developing its technology in-house and then running its own clinical trials, with little need for government participants. Moderna's, though, was the culmination of a five-year partnership between the National Institutes of Health and an upstart biotech enterprise that had neither the credibility nor the in-house scientific know-how to run its own trials. This had led to a model whereby the company designed the technology and government scientists like Barney Graham oversaw the experiments and clinical trials that would determine whether it worked. And while it may not have been what Aled Edwards envisioned in a public-facing path to drug discovery, the collaboration among Graham, Zaks, and Baric was nevertheless evidence that a highly engaged government, the competitive drive of the for-profit market, and the open-ended possibilities inherent to academic science could come together to help build a cure for the world. It was, for those who believed in the enduring need to support publicly funded science, a watershed moment. This made Graham's response to the news understandable: he sat in his office and sobbed.

ON NOVEMBER 23, 2020, THE VACCINE PARADE CONTINUED: AstraZeneca issued a press release saying that its adenovirus vaccine, AZD1222, containing a sturdy DNA fragment with instructions to build spike proteins, protected 70 percent of participants from COVID-19, but that it might reach 90 percent efficacy under certain conditions. In a strange admission, AstraZeneca explained that an error had caused some of the participants in its clinical trial to receive only half of the intended dose of the vaccine for their first shot. Stranger still, the group that was given the half dose appeared to be better protected. The errors notwithstanding, AZD1222 was rugged: its combination of an adenovirus shell (more rigid than the oily bubbles used to shield the Pfizer and Moderna vaccines) and a DNA fragment (less prone to disintegration than mRNA) meant that neither the vector nor the payload would break down easily. With such a hardy package, there was no need for a deep freeze as there was for some of its rivals; instead, AZD1222 would remain unspoiled as long as it was kept at 4 degrees Celsius, giving hope that it might be a global solution to supply far-flung places where cold chain distribution (keeping samples at minus 20 degrees Celsius, which is what the Moderna and Pfizer vaccines required) was next to impossible.

"Isn't it good news?" said Baric. It had been two years since he first linked up with Barney Graham and Tal Zaks to test CoV vaccines (first for MERS, and then for SARS-CoV-2) and six months since he raced to submit safety data on the effects of mRNA-1273 on mice in advance of human trials. The results clearly delighted him. As an elder statesman of coronavirology, Baric might have been expected to remain poker-faced. But the vaccine efficacy data were proof that his work, after decades of promise, would save lives. The "fine decision-making" to which he self-mockingly referred, and which had led him to launch a career probing the *Coronaviridae* instead of HIV, was no longer a joke. He had made a bet that pursuing an abstract question—why coronavirus genomes were so large— would eventually pay off. Forty years later, it had come to fruition.

Amid the celebration, though, Baric was still looking ahead to the next coronavirus. The results were only interim, he reminded himself, and there were still a few months during which adverse events could show up in the trials. Nevertheless, the data from his mouse models and from the early-stage human trials so far suggested that the vaccines were overwhelmingly safe. "Ultimately, if there is an adverse reaction, you may not see it until you've vaccinated five hundred thousand or a million people," he said, running numbers in his head. "But that might mean that the background of adverse reactions is one in a million or two in a million." This made the decision about whether to get vaccinated or risk natural immunity pretty straightforward. "The mortality rate of the virus is one percent. So, if you're flipping coins to decide what's best for your health, the answer is clear: take the vaccine. You're either going to get immune by natural infection, or you're going to get immune by the vaccine. And one looks relatively easy," Baric said, "and the other looks pretty rough. I don't know how else to say it." The added bonus for the mRNA vaccines, he pointed out, was that they could be quickly reformulated. Programming mRNA to encode different spike proteins took a matter of hours, and producing a vat took a few weeks, making the platform highly adaptable in case deadlier variants emerged. A combination of Baric's vision for a broad-based pan-coronavirus antiviral and a rapidly produced, reprogrammable vaccine pointed the way to a future in which humanity could prepare itself to fight back against the next pathogen.

THE ANNOUNCED EFFICACY OF THE VACCINES READ LIKE SIMPLE MATH: for every hundred people who got a shot, roughly ninety-five (in the case of Pfizer), ninety-four (Moderna), or ninety (AstraZeneca) would be protected. But it wasn't quite that simple. The crude numbers hid the stark reality that there were ways for coronaviruses to outrun our best weapons, reprogrammable though they may be. Since the publication of *The Value of All Chances in Games of Fortune* in 1657, scientists judged success or failure not by a specific number but by how much doubt they had that what they were seeing wasn't just random chance. Case in point: summing up the results of the Moderna vaccine trial pithily saying, using one figure, that it had 94.1 percent efficacy in protecting against COVID-19, was completely different from the vaccine's efficacy range of between 89.3 and 96.8 percent.

In Moderna's case, mRNA-1273's range was remarkably narrow

and staggeringly high. Even if you knew nothing about the study, an efficacy range of only 8 percent revealed that the study was likely conducted with a lot of participants and that the impact of the vaccine was almost uniformly positive. All that information was lost, though, in a single, lonely point estimate. Compare that to Astra-Zeneca's AZD1222: The vaccine was reportedly 70.4 percent effective, much higher than the 50 percent threshold set by the FDA, but the range estimate told a more nuanced story. AZD1222 was between 54.8 percent and 80.6 percent efficacious, a range roughly twice as wide as that for Moderna's vaccine. This wasn't necessarily cause for alarm, but it was a sign that the data told a more complicated story and that the vaccine was susceptible to dipping below the threshold of effectiveness if the virus continued to mutate. Graham, whose team had led the design of the synthetic spike protein, was unsurprised. AstraZeneca, he noted, was the only one of the major vaccine makers that hadn't opted to use the synthetic spike protein his lab had produced. "They're totally focused on the vector," he said, referring to the shell that managed to get the vaccine's payload into cells. "So, they just took the wild-type spike"—a spike protein the company had synthesized from the ancestral SARS-CoV-2—"and put it in their adenovirus vector." Graham was confident that AZD1222 would still work, but he cautioned that it wasn't as good as it could have been. "And I don't know if that's all because of the protein design, but it could be better."

MORE THAN 120,000 SARS-CoV-2 GENOMES HAD BEEN SEQUENCED BY December 2020, making up an army of poorly cloned pathogens, largely identical save for random atomic rearranging. Some of those mutations had, true to form, produced breakout variants capable of evading humanity's defenses. The *Coronaviridae* was an ancient family—as old as three hundred million years, by some estimates, far older than *Homo sapiens*—tricky, elusive, and full of guile. It was an enemy that knew us, and had, over millennia, honed its capacity to mutate to elude our best defenses.

Baric described the rise of SARS-CoV-2 variants as "competition for efficient transmitters," where the most efficient transmitter (first D614G and then the many variants that came after) was able to outcompete the others and become dominant. "That's what happens when you have, in essence, eight billion naïve people among whom maybe ten percent are immune," he explained. Whichever variant moved fastest got the prize: billions of hosts. Even if the family hadn't yet found the right combination of mutations to totally subjugate our species, Baric was sure its members possessed all the tools it would require to do so.

But knowing that SARS-CoV-2 might fragment into a multiplicity of variants and seeing it actually happen in real time were two totally different things. The news of variants emerging cut deep for Baric, who had worked around the clock to test vaccines, antivirals, and monoclonal antibodies against SARS-CoV-2; that work might well be for naught if the virus started undergoing extreme variation. All he could do was brace himself for the onslaught and hope that it was not as bad as he feared.

First to emerge was B.1.1.7, later dubbed Alpha, which was detected in early December 2020 from a patient in the county of Kent, in southeastern England. Though national restrictions had been established in the United Kingdom, and the incidence of new cases was falling or leveling off elsewhere, infections were rising in Kent. Public health measures like lockdowns and mask mandates hadn't changed, prompting virologists to start probing elsewhere to understand whether it was the virus itself that had adapted to the barriers humanity had placed in its path.

What they found was that Alpha was riddled with mutations. At seventeen different points in its genome, errors had been introduced that had transformed it into an even stealthier pathogen than D614G, the initial breakout variant that had superseded the ancestral SARS-CoV-2 strain and, early in the pandemic, taken over the world. But D614G had nothing on the multiple mutations that made Alpha a much-improved antagonist. These included N501Y, a swap of the N-amino acid for a Y-amino acid on the spike protein's receptor-binding domain that gave the variant the ability to grip on to cell walls more tightly. Once Alpha was flagged, scientists looked back at previously mapped genetic sequences and found that the variant had actually been circulating in the United Kingdom since September, four months earlier than the Kent index case. And while it initially comprised only 0.05 percent of all cases, within four months Alpha had fully crowded out D614G, which was now the laggard. Fully 90 percent of all infected people in the United Kingdom had the new variant, and some estimates suggested that it had caused as much as 50 percent more cases compared to its mutant ancestor. But while Alpha spread more rapidly, it also withered under the weapons (vaccines and monoclonal antibodies) that had allowed humanity, ever so cautiously, to dream of an end to COVID-19. Alpha was, thankfully, not going to postpone that dream. For all the advantages they provided the variant, Alpha's mutations were minor enough that monoclonal antibodies (which attached themselves to the spike protein, nullifying the virus's capacity to enter cells) still fit the spike protein's shape. Similarly, the antibodies generated by COVID-19 vaccines fit well enough that they could stop Alpha in its tracks.

Then came Beta. First detected in South Africa in October 2020, the new variant was ruthless. It was equipped not only with the D614G rigidity power-up and the N501Y tight-binding mod, but its spike protein carried another mutation, E484K (an E-amino acid swapped for a K-amino acid in the 484th link in the spike's molecular chain), which risked unraveling the advances humanity had made. E484K was an escape mutation, a lucky error that allowed the virus to transmit more efficiently while also evading human immunity. The mutation lay in an unremarkable, stringlike loop of molecules on the spike protein, bound together by electromagnetic forces, hydrogen bonds, and electrostatic energy. These forces competed and colluded to give the spike protein, in extreme close-up, a wispy, coiled look and primed it for its collision with a host's cell. E484K subtly shifted these forces, so that when the variant's spike bound itself to a cell wall, the bond was so tight that no immune weaponry could dislodge it. But there was more. The electrostatic charge caused by E484K deflected neutralizing antibodies away from the site where the virus met the cell wall, which repelled any antibodies sent to attack the virus. The result? The older variants were left in the dust, having quickly become relics, as Beta's emergence coincided with a sevenfold increase in COVID-19 cases in South Africa and the variant's spread to forty-eight countries worldwide.

Still, there were more. Gamma, a variant first detected in Japan on January 6, 2021, and traced back to Brazil, also had the E484K mutation. Its torrid spread across Brazil after a massive first wave triggered worries that this variant was so different from the ancestral SARS-CoV-2 virus that people could be reinfected with the deadlier descendant. The variant B.1.526 emerged in New York City in February 2021, sparking a surge in cases. There was also B.1.427, a variant first identified in California, and Fin-796H, in Finland. As the pandemic matured, setting off waves of infections across the world so numerous that they became as regular as the tides, the variants flourished.

And then there was the Delta variant. First detected in India in late 2020, Delta was the apotheosis of viral efficiency. It contained many of the mutations common to variants of concern, along with others that made it an even more refined predator. Key to its enhanced infectivity were spike protein mutations that dampened the immune system's capacity to recognize the virus as a threat; it also appeared to be able to separate the spike protein more quickly into its multiple components, which speeded up its entry into cells. The agglomeration of mutations had their desired effect: Delta was over 60 percent more transmissible than the original SARS-CoV-2 strain and largely evaded treatment from bamlanivimab, the monoclonal antibody that Davey Smith had tested in ACTIV-2 (bam's failure to work against Delta led the FDA to revoke its emergency-use authorization of the drug). The wild-type isolate first sequenced in January 2020 had become an ancient relative just one year later as its stronger, faster, and deadlier brood conquered the world.

With the steady uptick in COVID-19 survivors and the beginning of the vaccination drive, the human population was slowly accruing antibodies that targeted the SARS-CoV-2's spike protein and interfered with its capacity to infect us. These were the first tepid hints of herd immunity, though with just over eighty million confirmed cases by the end of 2020, it was a fraction of our species. This defensive posture had been enough to put selective pressure on the virus to adapt. While each variant was a wholly new creation of nature, collectively, in the face of humanity's efforts to mount a defense, they became something else: acts of viral revenge. The coronavirus was fighting back.

There was a part of Baric that felt something bordering on admiration for the pathogen he had spent his career studying. Its organic, unpredictable elasticity allowed it to find solutions out of apparent dead ends, cross the endless gulf that separated species, and overcome the immune resistance it encountered. What he was most worried about, though, was how the variants would adapt to the weapons humanity had belatedly brought to the fight. "The biggest concern is whether the Pfizer and the Moderna vaccines, for example, are going to still be effective against these variants," he said. The answer wasn't a simple binary of whether the vaccine worked or not. "The question is going to be the fringes." That's what had happened with SARS and was still happening, albeit slowly, with MERS: both viruses tended to punish a specific subset of those they infected. In any population that is either infected or vaccinated, Baric explained, the immune response follows a bell-shaped curve. "There's some people that have lower responses, and other people have fantastic responses, and most of us are right in the middle somewhere," he said. "These variants function at the edges, and so they're interested in converting ten or twenty percent of the population into susceptible hosts." Each successful mutation might knock down the body's immune response tenfold. Over time, these minor mutations accrued on variant spike proteins to the point that vaccine-generated antibodies fit their targets less and less well. For these fringe cases— that is, those people who already had trouble mounting an immune response to the virus (the elderly, those suffering from chronic conditions like diabetes, or people with autoimmune conditions, cancer, or anything else that suppressed their immune system)—the mutations could neuter the vaccines completely.

Still, there was cause for some cautious optimism. Though Delta had emerged as the deadliest and most elegant variant of concern, the Pfizer and Moderna mRNA vaccines appeared to be 88 percent effective against infection, while the AstraZeneca vaccine conferred 60 percent protection. And all of them still remained over 90 percent effective at preventing serious illness. Evidently, the mutations that dotted the variant spike proteins hadn't yet allowed the virus to escape the antibodies that the vaccines taught the human body to produce. One year out from the first cluster of viral pneumonia in Wuhan, the cures humanity had built were bending but not breaking in the face of their protean targets.

Baric harbored doubts, though. Most of the vaccine makers had reported on their vaccines' performance against variants using results from blood tests taken among a group of vaccinated people, which they then pooled together to calculate an average level of effectiveness. That simply wasn't good enough for Baric. "Taking a large block of serum [from multiple patients] and testing it and then saying 'Our vaccine works against variants' is truthful," he admitted. "But it's batched all the information, and it's not telling the whole story." Within the data might be a hidden truth: that the vaccines weren't working on variants infecting the fringe cases, but there was no way to know unless the pharma companies became more transparent about what was really going on behind the pooled numbers.

Looking to the future, Baric saw a scenario whereby the variants, endlessly recombining whole sections of their genomes and drifting through copying errors, would become more extreme as the pandemic entrenched itself among human populations. "So, let's say a thousand virus particles are required to infect someone now," he explained. "If the virus becomes more virulent, it could be that just one particle can infect an individual and initiate a robust infection." The other possibility was that the SARS-CoV-2 strain could recombine with an endemic human coronavirus like NL63 or OC43 through antigenic shift, swapping whole sections of genome; if that happened, SARS-CoV-2 might pick up spike protein components that made it more capable of evading our immune response and much more effective at spreading among humans. Baric marveled at the strategy relatively benign human coronaviruses had deployed to circulate across the entirety of our global population. "With the contemporary human coronaviruses like OC43 and NL63, the most vulnerable to severe disease infection are the very young and the very old, or those with comorbidities. Yes, fatal infections occur sometimes in young children, but they're rare. What happens most of the time is that as a young person, you're prone to infections—some kind of lower respiratory tract infection—but they're not life-threatening. And then every three years for the rest of your life, you get infected with common cold coronaviruses, the same or slightly variant strains, until you're seventy-plus. And then your immune system begins to wane, and then you can have severer infections in your lower respiratory system again, and that can be deadly. I think that's one logical scenario that the virus will use to maintain itself: it will infect children, and it will cause a mild common cold infection, where it can still be transmitted but it's not life-threatening."

There was also the chance that humanity's success against the virus might drive it into an animal reservoir from which it could then reemerge. Baric pointed out that the N501Y mutation first found in the Alpha variant, beyond making the virus more transmissible in humans, also produced infection in mouse and rodent populations, bat species, and other mammals. "Every time there's a change in one of those receptor-binding domains," he said, "the potential exists for a reservoir species to suddenly become prominent. And if it's something like a pig, or a cow, or a domesticated animal like a dog or a cat, we're going to have a problem."

This was the transmission cycle, Baric noted, that the virus had already established with minks. Indeed, in Denmark's remote North Jutland region, a COVID variant dubbed Cluster 5, detected on November 5, 2020, was found to have infected more than two hundred people. Cluster 5 was quickly traced back to mink farms, having likely arisen after a human worker infected the minks. The virus, after circulating in the animals, then mutated before spilling back out into humans. It was exactly what Marion Koopmans had feared would happen in the Netherlands and the reason she successfully pushed the Dutch government to ban the mink trade. Though it was unknown whether Cluster 5 weakened vaccine-induced immunity, early reports suggested that it was impervious to commercial monoclonal antibodies. Luckily though, Cluster 5 seemed to be an evolutionary dead end, poorly adapted to spread through human populations after adapting itself to minks. On November 19, two weeks after it was first detected, the Danish government announced that Cluster 5 had likely gone extinct. Along with the virus, seventeen million minks were killed.

There was a final fear that evidently troubled Baric. The trade-off hypothesis of viral evolution, he explained, held that highly transmissible viruses generally became less deadly over time because the host had to survive long enough to be able to transmit the virus to other people. It was a comforting thought, but Baric knew that it wasn't always the case. "There are obvious viruses that break that rule," he noted. "HIV is the best example, where it has close to one hundred percent lethality." The reason for that was the length of time between when an HIV-infected person developed symptoms and when they died of AIDS, which could take years, during which they could infect other people. SARS-CoV-2, though far less deadly, operated in the same way: there was an early asymptomatic period, during which an infected person had the highest levels of virus in their system; only later, when they started to develop symptoms, did the viral load go down. "And so," Baric said, "the scenario exists in a coronavirus where you could select for high transmissibility because most people who die from infection are dying at day ten to twenty, post-infection." In short, the trade-off hypothesis might not apply. "As far as I can tell," he continued, "coronaviruses may be capable of escaping this evolutionary constraint, in the same way that HIV has." The variants could make the coronavirus both more transmissible and deadlier, all at once. "My gut feeling is that it could happen. And it would be a horrible thing if it did."

Baric developed a natural self-effacing tone when he shared his profound knowledge of the *Coronaviridae* family and its many secrets. Perhaps it was born of the understanding of how short his brief foray into their universe was compared with their extended genealogy, which potentially stretched back farther than the beginning of our species and might just outlast us. Ever since SARS, twenty years earlier, Baric had been planning for a coronavirus pandemic. When one actually emerged in the form of SARS-CoV-2, even that didn't stop him from resolutely focusing on what dire threat it would pose next. "So, I guess there's a lot of things it could do," he said. "What it's going to do, I don't know." In the meantime, he would do what he always had: work harder to understand the viral forces that drove selection, extinction, and resurrection and try to end them.

"THERE ARE EVEN MORE INSIDIOUS CRISES WITH POTENTIAL FOR EVEN greater catastrophes than those posed by COVID-19." So intoned Malik Peiris, the virologist who first isolated SARS and helped identify the camel–MERS link, as SARS-CoV-2 fractured into variants that spilled across the globe, each new entrant accelerating past the last. It was an odd statement to make as the planet remained utterly gripped by the novel coronavirus, the third that Peiris had had to contend with in his career. But despite the trail of sickness and death that lay in its wake, Peiris was considering something even more terrifying: the possibility that SARS-CoV-2 would recombine with other deadly coronaviruses and spark the rise of another, far deadlier pathogen.

In the years leading up to the COVID-19 pandemic, Peiris had remained resolutely fixed on unraveling the ongoing mysteries of MERS, which by 2019 had been largely abandoned after the initial outbreak in 2012 had sparked a small flurry of research. Once the basic questions about MERS had been answered—most notably that camels were the likely reservoir from which it had jumped into humans—most virologists had moved on to more pressing concerns. But not Peiris. After his attempts to enter Saudi Arabia had been denied, he ended up sampling camels across the Gulf states and Africa—and what he found had confounded him ever since. Despite MERS being present everywhere he looked, Saudi Arabia and its neighbors had a much higher proportion of both asymptomatic and symptomatic MERS infections compared to African countries. Some of the severer cases could easily be ascribed to diet and lifestyle factors: wealthy, older, and obese Saudis (for example, the Bisha businessman who first succumbed to MERS) fell into the fringe cases that Baric had shown were more susceptible to coronavirus infection. But this didn't explain why there were fewer overall infections among people in African countries, where MERS-infected camels were ubiquitous.

Peiris wanted to go deeper, so he mapped the genetic sequences of a MERS virus harvested from camels in Africa and another one collected in Saudi Arabia. When he compared the two, he discovered subtle but important differences in their genomes that, while minor, pointed to different levels of transmissibility. "Africa is the birthplace of MERS," Peiris explained, "and the hypothesis is that there are strain differences that make the African virus less able to cause disease in humans." He was intent on finding out what was stopping MERS from spreading in Africa, but he ran into a problem that had plagued coronavirus research throughout the discipline's life span: lack of interest. "You'd be surprised," he said, "how many times I've put in grant applications to fund this work, and each time they say, 'Why are you wasting time looking in Africa? There's no problem there.'"

Then COVID hit, and Peiris found his obsession with MERS tested. As was the case each time a novel coronavirus arose that threatened the planet, he was called to action. His work ran the gamut from tracking how SARS-CoV-2 virus traveled through the air to identifying key sections of its genome that might be vulnerable to emerging therapies. But throughout, he never stopped thinking about MERS. "I don't have the time to write up the papers on MERS, because I'm now more or less ninety-nine-percent focused on COVID," he lamented. (In an email to Ian Lipkin, Peiris, in his understated way, described himself as committed to MERS but "a bit sidetracked by COVID-19.") "But MERS is extremely interesting, and I wouldn't be surprised if, God forbid—I don't think we can tolerate anymore right now; I certainly can't tolerate another pandemic like this," he said, "but I wouldn't be surprised if MERS comes and hits us while we are still in the middle of COVID."

Peiris's fear was grounded in those subtle mutations that separated African MERS virus, which had proven largely incapable of spilling over from camels into humans, from the Saudi MERS virus, which had found a way to infect humans and kill over a third of those who became ill. In virology, subtle differences are simply a matter of time: the fewer chance mutations that separate an innocuous virus from a deadly pathogen, the sooner the transformation might occur. "If that change could happen once," Peiris said, referring to the mutations in the Saudi MERS virus that separated it from its more benign African progenitor, "then it could happen again. And if that happens in a wider area of Africa," he continued, "you would then have a whole different ball game in regard to MERS. There would be much larger populations infected and the potential of a pandemic."

As SARS-CoV-2 variants spread across the world, the fact that other coronaviruses were also constantly mutating had largely been ignored. And yet, the pandemic provided a short window of time for other pathogenic coronaviruses to jump-start their adaptation to our species. As Baric well knew, the vast spread of SARS-CoV-2 opened up the possibility that it might recombine with other endemic and innocuous human coronaviruses like OC43 and NL63. That might make it more transmissible, less deadly, or both. But the very real possibility also existed that SARS-CoV-2 could recombine with MERS, its far deadlier cousin, and impart to the slow-moving pathogen the capacity to transmit itself efficiently. The upshot would be a MERS/SARS-CoV-2 hybrid that killed upward of a third of those it infected and that could transmit itself rapidly and asymptomatically across large swathes of the global human population.

When Peiris was in junior high school in Sri Lanka, he read a biography of the life of Louis Pasteur, one of the discoverers of vaccines, by the author René Dubos. The book contained a quote that stuck with him and helped spur his eventual path to becoming a virologist: "At some unpredictable time and in some unforeseeable manner, nature will strike back." While the world at large believed that COVID-19 was the extent of nature's wrath, Peiris, a coronavirus veteran, was bracing himself for the violence to come.

Chapter 15

WE LIVE IN A GRAY WORLD

November 20, 2020, Chapel Hill, North Carolina

THE VACCINES WERE POTENT NEW WEAPONS TO STANCH THE spread of SARS-CoV-2. But in a strange confluence of timing, the very same week in which the interim results of the major vaccine trials were announced to great fanfare, Ralph Baric's antiviral arsenal was beset by a stunning reversal.

It had been seven years since Baric and Mark Denison hatched their pandemic-busting plan: to test and stockpile broad-based antiviral drugs that could work against any coronaviruses, past, present, and future. Moving through a database of three hundred thousand drugs, they had eventually winnowed the list down to just two, remdesivir and EIDD-2801, both of which they had tested exhaustively and found to be effective at blunting coronavirus infection and disease. Tired of waiting for Merck to launch a clinical trial of EIDD-2801—it eventually did in October 2020—Baric had helped launch his own the previous May, with results due by early summer 2021.

The same wasn't the case for remdesivir, which had been put through its paces in human trials in the early 2010s by the pharma giant Gilead as a would-be hepatitis therapy and then as an Ebola drug. Though it failed in both instances, at least there was more than enough evidence that it was safe to use in humans. This head start set it up to become the first therapeutic drug approved for use against COVID-19. (Others, including the monoclonal antibody bamlanivimab, which Davey Smith had tested in ACTIV-2, later joined its ranks.) Since its initial authorization, remdesivir had become a billion-dollar drug for Gilead. It was proof, Baric said, of just how lucrative planning for future pandemics could be and a good reason to bet on initiatives like READDI that were actively preparing for the next one.

Though remdesivir had become the first approved treatment for people hospitalized with COVID-19, this hadn't stopped the grinding gears of scientific inquiry from churning. In June 2020, the WHO had launched an international drug trial dubbed SOLIDARITY, aiming to test whether remdesivir (among other therapies) reduced death among inpatients. It was a large, well-funded study and appeared to hit the benchmarks for study quality, such as randomizing patients to receive treatments and recruiting enough of them (roughly three thousand) to ensure that the results the study generated were high in confidence and low in doubt. Baric, though he wasn't involved with the study, had been following SOLIDARITY closely and was hoping that the trial would provide remdesivir with that most worthwhile scientific currency: replicability, which would demonstrate that the first National Institutes of Health study of the drug, which had found a benefit, wasn't just a fluke. (Replicability was the same thing Tal Zaks desired, and received, when the Moderna and Pfizer mRNA vaccine trials produced similar levels of protection against COVID-19.) In that first trial, hospitalized patients who were given remdesivir were about 25 percent less likely to die. This had paved the way for remdesivir to become the first therapeutic agent authorized by the U.S. FDA to treat COVID-19, and this was what those who believed in the drug, like Baric, hoped SOLIDARITY would find as well.

SOLIDARITY did no such thing. Instead, the study found that when remdesivir was given to hospitalized patients, it completely failed to keep them alive. This was no statistical quirk or a biased result from a poorly conceived observational study. The SOLIDAR-ITY trial was about five times the size of the earlier NIH study, giving it more statistical power to detect significant differences between those who did and did not get remdesivir (though Baric cautioned that it used an open-label design, which allowed patients and doctors to know whether they were receiving the drug, which could have skewed results if it affected other aspects of their treatment). The reason for the study's discrepancy with the earlier NIH study, the SOLIDARITY scientists suggested, was that the initial NIH-funded trial had given remdesivir to hospitalized patients who were, on the whole, less severely ill than those who got a placebo. In short, the WHO accused the NIH-funded scientists of biasing their own study in favor of the drug by giving it to patients who were more likely to recover even without it.

On November 20, 2020, the World Health Organization released a blunt statement: "WHO Guideline Development Group advises against use of remdesivir for COVID-19." The announcement left no doubt as to what the organization thought about the drug Baric had plucked from hundreds of thousands of others to protect humanity. "The panel acknowledged that the certainty of evidence is low," the statement read, but that "there is no evidence based on currently available data that it does improve important patient outcomes."

The WHO's decision evidently weighed heavily on Baric, who for the most part appeared sanguine as he encountered the vagaries of life as a coronavirus researcher. But it was clear that the whole episode stung him deeply. Baric dismissed the results as predicated on a total misunderstanding of how remdesivir worked—and, more important, when it worked best. SARS-CoV-2, Baric explained, exploded early in its course of infection, with viral copies at their highest in the first few days, which then waned over time. "The virus goes up, and then it comes down," said Baric, "and then the immune system response comes up behind that and peaks almost after the virus has pretty much gone." The upshot was that by the time somebody with COVID-19 was ill enough to be admitted to a hospital, it was already too late: they had entered the second part of the biphasic COVID illness, in which the immune system was no longer fighting the virus but its own body. But remdesivir, the antiviral, worked only when the virus was saturating cells and exploding across the body, which happened in the first few days after infection. By the time an infected person was hospitalized, it was already too late. Baric figured that should be obvious. But nobody, he fumed, understood.

When the World Health Organization stamped its imprimatur of failure on remdesivir, it knocked the only antiviral drug off a very short list of COVID-19 therapies approved by the FDA. (That list also included the steroid dexamethasone; convalescent plasma from recovered patients; and three monoclonal antibodies, including bamlanivimab, which had been run through ACTIV-2 along with the anti-inflammatory drug baricitinib.) But aside from shrinking even further the paltry set of drugs available to treat COVID-19, the WHO's recommendation was a harsh blow for Baric's "one drug, many bugs" plan to end pandemics for good.

He tried to stifle his anger. Even so, an evident melancholy suffused his words. He had watched with skepticism as the legion of scientific experts who had come out of the woodwork during the pandemic extemporized about remdesivir's shortcomings. "Everybody who goes on TV tries to give an explanation," he said. "But these issues are difficult to explain on TV. People want to hear 'it works' or 'it doesn't work.' They don't want to hear that we live in a gray world." CNN, NBC, Fox News—all the major networks covered the story, with reporters simply repeating the standard line that the WHO study found that the drug was ineffective. Baric knew better: the issue with remdesivir was as gray as it got.

What frustrated him the most about the "new" science around remdesivir was how unnecessary it was. "The strengths and weaknesses of remdesivir have been known since we tested it in an animal model," he said, referring to results that he and Mark Denison had published way back in 2017, after testing the drug on fourteen different coronaviruses Baric had collected in his lab. "That data showed that it worked great as protection before infection, and it worked great early in infection. But in the mouse model, after about two or three days after infection, it completely fails." The SOLIDARITY trial was built on a misunderstanding about when remdesivir should be given to infected patients, which was as early as possible. Still, the fact that the antiviral could be injected only by healthcare professionals in a hospital setting gave scientists little choice in how to test the drug's effects. In the real world, patients were getting it after they were sick enough to go to the hospital, though this was a little like putting a seat belt on after a car crash. This stuck remdesivir in a catch-22: It was a drug that needed to be injected into the body, which could be done only within the walls of a hospital. But hospitals wouldn't admit anyone until their infection was so severe that they were sick enough to warrant inpatient care. But by that time, patients would already have passed into the second phase of COVID disease, and the drug wouldn't work anymore. The SOLIDARITY trial, as methodologically bulletproof as it might have been, was testing a drug at the wrong time.

"So, did the WHO make the right decision?" Baric asked with a sigh. "Well, if they're only going to use the drug at very late, end-stage cases, then it's probably not going to be effective, and it's probably the right call. Does that mean that the drug's not effective? No." Ultimately, Baric had no issue with the WHO's recommendations per se; after all, they were the only possible conclusions to make from the study. What incensed him was that the scientific paradigm was totally skewed: nobody evaluating the drug seemed to understand how it worked. But what could he do? He wasn't asked to consult on the trial, perhaps because his early work developing remdesivir would have given him the appearance of bias.

In any case, Baric was already eyeing the next pandemic. "The key is that Gilead," which owns remdesivir, "needs to develop an orally administered version." Baric could only hope that the bad press wouldn't lead to its being cast aside indefinitely. With the *Coronaviridae* all but certain to produce a pathogenic heir apparent, humanity would need to have every weapon at its disposal ready for the next battle.

CORONAVIRUSES EXPLOITED MULTIPLE PATHWAYS TO DOMINATE THEIR hosts, making them elusive targets capable of evading immune responses and antiviral weapons. Some experts believed that during their long coevolution with our species and others, the family even incorporated genetic material from the DNA of their hosts into their own RNA to improve their fitness. It was evidence of just how opportunistic the family could be. Baric had evidently learned from this technique, having stolen something from the family as well: the knowledge that one path was never enough. So, while the disappointment of remdesivir stung, he and Denison had long burnished another weapon: EIDD-2801.

Like remdesivir, EIDD-2801 could pass through cell walls easily and latch on to the coronavirus copy machine while camouflaging itself from the viral proofreader. But there was a key difference. While remdesivir had to be injected in a hospital, EIDD-2801 could be taken orally, anywhere and by anyone. And this made it the drug remdesivir might have been: a pan-coronavirus antiviral that could be easily distributed across a newly infected population to rapidly extinguish an outbreak before it expanded.

EIDD-2801 hit the same marks as remdesivir and then some. The relative ease of stockpiling and distributing a pill compared to an injectable drug made it a major plus. But EIDD-2801 had also proven itself more adaptable to the protean SARS-CoV-2 than remdesivir. In 2018, Baric had found that the mouse coronavirus MHV, the SARS-1 virus, and MERS could all develop resistance to remdesivir. But when he ran the same tests on EIDD-2801, the viruses were unable to escape: the antiviral drug conferred long-lasting protection even forty-eight hours after infection, and was even more effective when given prophylactically twelve hours prior to infection. Those results, released in March 2021, were something of a coming-out party for READDI. Baric, who published the findings in *Nature*, proudly listed the initiative as one of his affiliations, cementing it at the forefront of coronavirus science.

There was even more good news. When SARS-CoV-2 variants of concern started to emerge, a group in Glasgow led by the virologist Meredith Stewart decided to conduct a head-to-head comparison of remdesivir and EIDD-2801 to see how each fared in the face of the constantly mutating virus. Stewart, who was deeply inspired by Baric's work, introduced the progenitor strain of SARS-CoV-2 into human cell tissue and then added either remdesivir or EIDD-2801. She then let the virus grow, harvested it, and added it to a new cell line (a process called serial passaging), to coax it to mutate under different conditions. After the virus was passaged a mere thirteen times through cell culture, a mutation arose in nsp12 (the viral copy machine) that significantly knocked down remdesivir's ability to stop viral replication.

But EIDD-2801 was a different story. Stewart found that while remdesivir became essentially useless, the mutations in the virus's copy machine weren't enough to overcome the sustained bombardment of errors the drug wrought. The results were just the latest evidence that Baric, across a career frustrated by disinterest, government-mandated pauses, and misunderstanding, but marked by a singular capacity to make the invisible universe of the virome visible, had found in EIDD-2801 what he had been searching for: his "one drug, many bugs" solution.

With EIDD-2801, Baric's journey through coronaviruses proved neatly elliptical. The question of how the family maintained its impossibly large genomes was what had attracted him to the field in the first place. As they transformed into threats to humanity, his question shifted to finding the target that could stop their spread. Both pursuits ended in the same place: the viral proofreader. Unlike the coronavirus receptor-binding domain, which mutated at three times the rate of other sections of the viral genome, the proofreader was largely "conserved," a part of the virus that the entire family had in common, like a shared congenital defect or the same brown eyes. From NL63, which likely tormented humanity in the eleventh century and by the twentieth century had become merely a nuisance, to SARS, MERS, SARS-CoV-2, and the potentially millions of bat coronavirus variants that existed just beyond the reach of humanity—all of them had the proofreader. This made Baric confident that whatever coronavirus next emerged from the tangled viral mass that lurked just beyond the event horizon of our species would be similarly unable to detect EIDD-2801, which hid from the proofreader's sensitive touch and furtively inserted mutation after mutation until the whole edifice collapsed. Though he had been working toward it for years, it was nevertheless unfathomable. Baric, along with close scientific collaborators like Mark Denison and the ecosystem of biotech start-ups, government funders, and university labs, had landed on their once-and-future pandemic weapon.

Merck, the company that had licensed EIDD-2801 under the trade name molnupiravir, was evidently among the believers. This was why, after signing the drug up to be tested via ACTIV-2, Davey Smith's adaptive trial, the company had ultimately pulled it from contention. Presumably, Merck figured it could run a trial faster on its own, not to mention that it wouldn't have to reveal just how special the drug was to their competitors testing their own drugs through Smith's trial.

While Merck negotiated its way to a human trial, Baric was thinking about more than just EIDD-2801's potential as a medicine to give to sick people. Envisioning a future pandemic that tore more voraciously through human hosts, he knew that it would not be enough simply to treat people who were already infected. The most effective way to stop the sickness would be to dose people—most likely, frontline healthcare workers—before they caught the virus and to do it early in the pandemic. This approach, perfected in combating the HIV pandemic, was known as pre-exposure prophylaxis, because it involved identifying people at risk of getting infected before they were exposed. It was also akin to "ring vaccination," a strategy successfully used to fight Ebola in which uninfected people close to an infected index patient were vaccinated first so that the virus had no paths to travel out into the broader population. The only differences were that EIDD-2801 was an antiviral and it could, in principle, be used to battle not only SARS-CoV-2 but also what-ever grotesque progeny the *Coronaviridae* produced next.

But the drug's great promise didn't make it a silver bullet. The modern-day history of epidemics was a story less about scientific discovery than about a deep imbalance in resources. In 1996, the Canadian HIV scientist Josef Decosas remarked that if the cure for AIDS were a clean glass of water, over half the world's population would still not have access to it. Decosas's point was that innovation, while important, ran low down on the list of reasons that deter-mined whether people in the midst of epidemics lived or died. Baric was keenly aware of this problem and had applied it to his thinking about the future of SARS-CoV-2. "Long term," he explained, "I think it will eventually settle down and become sort of a mild com-mon cold infection in adults," just as NL63 and OC43 were believed to have done. But there was one barrier keeping us from that place of balance with the virus: "We've got to get to sufficient levels of immunity," Baric said. "And for most of the globe, it's going to be natural infection, sadly."

Baric had no doubt about where across the globe this natural infection would arise and what it would look like. "The waves of disease will be more serious, and longer-lasting, in the developing world than in the U.S., and Europe and those countries that have developed and purchased the vaccine." It was as simple as that. Wealthy people would be protected from the pandemic; poor people would not.

But there was more. Though this morally bankrupt situation was the global status quo before the pandemic, SARS-CoV-2 had turned it on its head. Baric knew how skillfully the virus took advantage of whatever hosts it had access to in order to evolve further. If left to recombine in endless transmission cycles, whether in minks or among unvaccinated people in poor countries, it would be just a matter of time before new variants emerged that could elude the vaccines. And when that happened, it wouldn't be only unvaccinated people at risk. It would be the entire world.

The vaccine economy was out of Baric's control. But like HIV, which can be controlled by a cocktail of antiretroviral therapies, EIDD-2801 had shown that it could disrupt not only SARS-CoV-2 but all other coronaviruses that had yet to, or would likely, emerge. Under the onslaught of variants, the drug maintained its capacity to destroy. But even so, Baric knew that this wouldn't mean a thing if the world failed to move on his findings or, even worse, if rich countries hoarded EIDD-2801 and poor countries were denied it.

Baric had done his part. He and Mark Denison had led humanity by the hand toward an antiviral that could end coronavirus pandemics forever. Even so, it wasn't enough. The weapons mattered only if they were purchased and placed in everyone's hands.

"RALPH'S AN EXTRAORDINARY SCIENTIST, BUT BY NO FAULT OF HIS OWN, he's never been involved in making a therapeutic or a vaccine, other than a cog in the continuum." Aled Edwards was in a state and, true to form, opening fire on friend and foe alike. In READDI (his plan to develop antiviral therapies attuned to future viral threats from coronaviruses, alphaviruses, and flaviviruses), Edwards saw a way to take the world's sudden fear of coronaviruses and use it to create an open-science platform to develop drugs outside the cruel strictures of the capitalist system. The pandemic had scrambled the prevailing notion that there was no way to design drugs that didn't directly feed into the market for them in wealthy countries. But READDI was achieving something unheard of: it was developing drugs for pathogens that didn't yet exist.

As the conflagration of variants was proving, the only way to stop pandemics was to stop them everywhere, in rich and poor countries alike; failing to do so threatened everyone, as it gave the virus pockets where it could continue to evolve. If READDI were going to actually make a difference, Edwards believed, it would need to reject the norms of scientific moneymaking. In practice, this meant instituting his cardinal rule: no patents.

When the former Eli Lilly marketing executive John Bamforth joined UNC and took on oversight of READDI, he brought a savvy marketing perspective and an aggressive goal-oriented leadership style born of his time running a $6 billion pharma marketing division. Edwards, who had also invited Bamforth on as a board member of the Structural Genomics Consortium, loved all that. But when the topic of their meetings started shifting from how to structure READDI to how to pay for it, cracks began to appear. Edwards was supremely confident that READDI could entice deep-pocketed funders without resorting to the sordid topic of patents. This had worked with the Structural Genomics Consortium, after all: going on almost ten years, the consortium was a $250 million open-science enterprise, bankrolled primarily by pharma money, which totally eschewed patenting the proteins it mapped or providing its patrons with anything other than an essential service that benefited the collective. READDI was, Edwards assumed, an extension of that model. And anyway, in the midst of COVID-19, why would anybody require convincing about the need to prepare for future pandemics? The future threat that Baric had been warning humanity about had landed; it was real, and nobody wanted a repeat. COVID-19, Edwards figured, was a painful lesson in the benefits of planning ahead. With the next pandemic a when, not an if, there was no need to sweeten the pot with patents: simply being part of READDI would be its own reward.

This was no idle thought. Remdesivir, which Baric believed would still be effective against future coronaviruses despite wavering doubts about its effectiveness, was priced by Gilead at $3,120 per dose. With the drug being one of the few authorized treatments for COVID-19, the price tag hadn't stopped clinicians across wealthy countries from prescribing it. Gilead had, on the back of Baric's research, turned it into a billion-dollar moneymaker. But it had also in the process made it entirely unavailable for vast swathes of the world. If it weren't for Baric's publicly funded research, Gilead would still be stuck with a failed antiviral collecting dust on a shelf. Baric figured this financial incentive was a great way to attract other drugmakers to READDI. Case in point was EIDD-2801, or molnupiravir, the little brown pill that could be orally administered and was less likely to produce drug-resistant coronavirus strains compared to remdesivir; one of the chief reasons that Merck had licensed it was because of the evidence Baric and Denison had generated showing it worked on multiple coronaviruses. But Edwards saw that profit motive as a profound problem. He didn't want the next remdesivir to be left to the private market, where it simply didn't exist for poor people. He was determined not to continue that legacy of moral failure.

Edwards could rail against patents all he wanted, but the way READDI had been put together left him on the sidelines. He led the Structural Genomics Consortium from Toronto; READDI was based out of the University of North Carolina at Chapel Hill. It was ultimately up to John Bamforth, who oversaw READDI, to decide how the organization would attract funders to the project while also keeping UNC happy. And that, Edwards claimed, was where the problems began. UNC, like every other academic institution, wanted to be "innovative," a word that sounded to Edwards like nails on a chalkboard. Innovation, he knew, led to patents, which led to greed, which led to people in poor countries being unable to afford the discoveries the universities made. Innovation was the snake eating its own tail. With this in mind, Edwards found Bamforth's title at UNC almost laughably triggering. He was the director of the Eshelman Institute for Innovation, tasked with doing exactly what Edwards hated: monetizing science.

Though Bamforth had ostensibly agreed with Edwards that READDI should be an open-science project, Edwards had grown alarmed as the former's resolve wavered in the face of pressure from inside UNC. Edwards knew from experience that as soon as universities took an interest in a project, they would look to make money off it. What's more, Edwards claimed that Baric, like so many scientists who had only ever worked within universities, didn't sufficiently understand the insidious way that academic institutions used patents to fund themselves and keep medicines expensive and inaccessible. "Ralph and John will just go along, because they don't understand that if you're trying to effect change, you can't make compromises.

And once you compromise—" Edwards stopped himself. "Actually, *compromise* is the wrong word. It's that you have principles, and you just stick to them." In their recent conversations, Bamforth seemed to be wavering about keeping the future-focused antivirals READDI had discovered patent-free; there was just too much pressure from UNC brass to make some money off the venture. "I'm beyond frustrated," Edwards seethed. READDI was worthless if the drugs it developed were too expensive to be stockpiled in rich and poor countries alike in advance of the next pandemic. Already, Baric's work on remdesivir had shown what could happen when you weren't careful: a drug could be brought to market that most of the world couldn't afford. It was no coincidence that UNC's long-standing support of Baric's work had allowed the university to profit from the success of his lab. "This is very, very typical," Edwards said. "People who don't start with the logic flow—their incentives at UNC are not to create a better world, their incentive at UNC is to enrich UNC and only then create a better world."

In Edwards's view, Baric had also failed to recognize that his work on remdesivir and EIDD-2801 gave him leverage over his pharmaceutical partners, and he should have used that to push them to make the drugs widely accessible. "I mean, Ralph—because he's been exposed only to the closed side of science, of pharma, he just thinks that's how it works. And then anyone who doesn't think like that is naïve."

Compounding matters, READDI was among the highest-profile projects that the University of North Carolina had going during the pandemic, but the initiative still didn't have any major funding, giving it little leverage. If the university were going to support it, it would need to get something in return. And if the university got something, then every single prospective funder would feel entitled to something, too. "READDI won't be able to find collaborators," he predicted, "because as soon as you have closed science and patents, you have to have legal agreements that encumber you. Collaborators will look at that and say, 'UNC is looking for barter money.'" If he were doing the negotiating, Edwards said, it would be an open-and-shut case. But with Bamforth in the driver's seat, after thirty years of perfecting the art of pharma profiteering, the prospects were dismal. "What John is going to do is say, 'Oh, we can accommodate open science, and closed science, and proprietary science,' right, and that's not going to help."

Edwards admitted that none of this was a surprise. After all, he had invited Bamforth to be a board member of the Structural Genomics Consortium fully aware of his background in sales. "Pharma companies are really two companies in one," Edwards explained. "The research-and-development guys are like me. It's just that when they were making career decisions, they asked themselves, 'Do I want to do something esoteric or something practical?' The ones who answer 'practical' go into pharma. And so, their hearts are the same, their ethos is the same; they just made that personal decision: esoteric, practical." Not so those on the other side of the industry, like Bamforth. "The sales guys," Edwards fumed, "are odious."

Edwards, reading the tea leaves, had made a decision. If Bamforth was going to be weak-kneed and sacrifice the young enterprise to the ancient forces of institutional greed, then Edwards was done with READDI. Developing antivirals to stop future pandemics was vitally important, but only if they were available to all. READDI was doomed to fail if left in Bamforth's hands. "We need to prove by example," Edwards said, by building a fully open-science approach that won't leave poor countries on the outside looking in.

Bamforth barely batted an eye as Edwards's ire rained down on him. Since joining the board of the Structural Genomics Consortium in 2019, he had become inured to the barrage of criticism of patents, the private sector, and pharma that emanated from Edwards. Bamforth claimed that those things didn't bother him very much. "So, Al and I, we're not one hundred percent aligned," he said with characteristic understatement. "I always joke with him, but the reality is that the Structural Genomics Consortium, the organization that he's the CEO of, is funded by pharma companies creating intellectual property from what his team discovers through an open-science lens. That is just a statement of fact." Edwards could rail about the dangers of patents all he wanted. That didn't change the fact that the consortium he so proudly led existed at the pleasure of pharma companies to help them make higher profits. Edwards called Baric a cog in the continuum, but Edwards, Bamforth was suggesting, was a cog himself, keeping the engine of commerce moving.

Bamforth was sanguine. "I spent thirty years in the pharma industry while he spent thirty years in academia. It's not too surprising that we have slightly different views on this thing. But I see a lot more examples of pharma companies delivering products by creating intellectual property than I see products being developed in complete open science all the way to the marketplace." Bamforth went further: he hadn't seen a single drug make its way to market without a patent. Edwards could rail against market forces until he was blue in the face, but it wouldn't change the reality that his purported "better way" didn't exist.

Baric refused to be drawn into the fight. "The basic question is," he said, "are you going to take a proactive or a reactive response to the pandemic?" Getting treatments to low- and middle-income countries was going to be difficult no matter what business model READDI used. In the meantime, there was a more pressing need. "The only way to take a proactive role," Baric said, "is to identify a drug that works against every coronavirus." With SARS-CoV-2 able to develop resistance to remdesivir, the world had only a single broad-based antiviral that could, in principle, protect our species from present and future pathogenic coronaviruses. Baric didn't like those odds. He had spent two-thirds of his life studying the *Coronaviridae* and knew how deeply adaptable the family was. Coronaviruses predated the rise of our own species, making us the newcomers; and yet, they had found a way to conquer our biology over and over again. In the face of that winning streak, Baric was leery of relying on just one drug, EIDD-2801, to protect us. The work of READDI had to continue at all costs.

Chapter 16

IT'S JUST TOO HARD TO SWIM
AGAINST THE TIDE

May 2021, Seattle, Washington

FROM HIS VANTAGE POINT WITHIN THE ICU, NICK MARK HAD SEEN the change in seasons since the first COVID-19 cases hit Seattle a year ago. He did not watch the gloomy sheets of cold rain that marked winter turn to the spring showers feeding the region's temperate rain forest and cedar giants, or the impossibly long summer days of the Pacific Northwest get whittled away into the brief daylight hours that marked fall days. No. Nick Mark's seasons were of a different quality altogether. The first pandemic season was one of chaos and fear as a flood of COVID-19 patients arrived in his ICU and lingered there for weeks, far longer than the brief few days most patient visits lasted. Draped in PPE, Mark and his colleagues did all they could; in the absence of trials on therapeutics, they experimented with any drug that sounded promising (hydroxychloroquine, azithromycin, baricitinib, convalescent plasma, remdesivir, dexamethasone) in the hope that one of them would hit. The second pandemic season saw the elderly replaced by the young and his own colleagues ending up intubated in the ICU, while Mark's adrenalized fear of the virus became a chronic condition he struggled to control. Farther afield, he did all he could to help those in other ICUs. In May 2020, via video from

Seattle, he guided junior medical residents in New York City who had been thrust into caring for COVID-19 patients in ICUs at risk of being completely overrun, his own despair barely under wraps. The third pandemic season, though, saw a change: with large research trials completed and Mark's own months-long experiences to draw on, the scattershot approach to COVID-19 care gradually became more refined, the drugs more reliable, and the symptoms more predictable.

The latest season of the pandemic had, finally, brought solace. On December 19, 2020, Dr. Nick Mark was vaccinated. The doctor was happy; his parents were overjoyed. After the initial thrill, though, he wasn't sure how to feel. Ultimately, his clinical instincts for playing probabilities were too ingrained to make him overly enthusiastic. "I have enough PPE at work, and I've been dealing with this virus for ten months, and I haven't gotten it, so it's not like this changes everything for me." Mark was still going to work, still wearing a mask, and still keeping his distance from other people. Aside from providing a little peace of mind, he concluded that the vaccine wasn't actually going to change anything about the way he cared for the victims of COVID-19.

Mark knew that he'd be in for many more pandemic seasons. "A year from now," he predicted, "the people with COVID are probably going to be those who got the vaccine and got sick anyway—the breakthrough cases." It'll be "people who are homeless, illegal immigrants who don't have great access to care, and people who are just uninsured. And then we're probably going to see people who are mentally ill, who are just not able to engage with the healthcare system, even if they are on disability." Aside from those vulnerable populations, there was one other group Mark was sure to see: those ideologically opposed to the vaccine. Their numbers would soon go up in the ICU. "I have a feeling a year from now, I'm going to be treating people who are, like, very into natural cures and stuff and are talking about vitamins as a way to stop COVID."

———

RALPH BARIC PRIDED HIMSELF ON CHOOSING COLLEAGUES WHOM HE also considered friends. It had paid off handsomely. His career was marked as much by discovery as by the long-standing collaborations that made his work possible. He had an evident fondness for the ecosystem of lab colleagues, partners, funders, and scientific acquaintances with whom he had crossed paths in his many years of coronavirology. The feeling was evidently mutual. Baric's admirers were near and far, but all spoke of him with real admiration—a titan, a giant, a prince of a human being, a close friend (even Edwards admitted to his brilliance)—and revered his body of work as awe-inspiring. It was easy to understand why.

But goodwill wasn't enough to convince the world to develop an antipandemic strategy that recognized that COVID-19 was only the latest, and certainly not the last, in a long line of assaults on our species by the *Coronaviridae*. For that, Baric needed READDI and the arsenal of antivirals it could develop and then deploy as soon as the next coronavirus outbreak flared.

But as the fall of 2020 turned into winter and as 2020 turned to 2021, READDI's success seemed far from assured. John Bamforth, one of the most successful marketing executives in the history of pharma, still hadn't locked down any of the hundreds of millions in funding he believed READDI needed to succeed. Worse, Aled Edwards's November 2020 tantrum had threatened to derail the effort and had left the group without his clear-eyed and ruthless strategic mind. As important, without the Structural Genomics Consortium that Edwards led, READDI would lose access to two hundred scientists around the world who could help it in its search for antiviral weapons.

Since the late 1990s, when Edwards first founded a biotech company and saw how patents were weaponized to restrict access to life-saving medicines, he had committed himself to creating a different path to discovery. "I was Derrick Rossi twenty years ago," Edwards said, referring to the Moderna founder's full-throated embrace of his scientist-entrepreneur persona and the profits that came with it. "I started my company because I wanted to make a difference—to make a milestone drug. But then I thought, 'Well, it's going to be priced at levels that will exclude most of the world, and then what?'" Edwards said. "That whole process made me realize that there has to be another way. And I've spent twenty years trying to make it happen. It's a fucking hard thing," he said, "because the Rossis are all over. And they say, 'Oh, eventually the drugs we make will reach poor children, but if we didn't do it this way, there would be no medicine at all.'" Edwards, liable to riling himself up, was riled. "Which sounds great. But then they say, 'Ah, but it's only after the patents expire in twenty years—only then will children in poor countries have it.'" For Edwards, this was no solution at all: delay equaled death.

(For his part, Derrick Rossi acknowledged the harsh realities of a for-profit approach—chiefly, the inevitability that the private sector would avoid producing cures for diseases in poorer countries where the market incentive did not exist—but believed that it was just the way the world worked. "I've heard the argument that it's not moral, that the rich countries are lining up, and they purchase at the expense of countries with less resources. I actually would argue that it's going to work out, and it is the only way it would work out because otherwise, how do those nations that don't have the financial resources pay for a product from for-profit companies to vaccinate their people?")

By the beginning of May 2021, Edwards had largely been proven right. Though the leaders of wealthy countries had loudly committed themselves to ensuring that vaccines were distributed equally across the world, the pious political talk and funding commitments hadn't translated into shots in arms in most low- and middle-income countries. The vaccine initiative dubbed COVAX (for COVID-19 Vaccines Global Access) had instead been an abject failure. Of the 2.1 billion vaccine doses administered worldwide, only 4 percent had been delivered via COVAX, while 75 percent of all vaccines had gone to only ten wealthy countries. "This is not only manifestly unjust," explained António Guterres, the UN secretary-general, "but it is also self-defeating," because SARS-CoV-2 would continue to evolve in poorer countries, threatening the effectiveness of vaccines everywhere.

This sorry state of affairs had once again proven to Edwards that it wasn't enough for scientists to produce a medicine that could save lives. They also had an ethical responsibility to ensure that the bargains they made in developing a medicine didn't end up excluding people from having access to it. "There's a lot of forces at work against that concept. And it's a lot of work pushing against something that you know is wrong but everyone else wants to do."

Edwards had been sure that Bamforth would succumb to the pressure. But to his surprise and delight, the ex–pharma executive proved him wrong: Bamforth had remained steadfastly committed to keeping patents off the table in his funding negotiations with UNC and other funders interested in supporting READDI. "My concern," Edwards said, "was that he would be flooded in the tsunami of pressure and just say, 'Okay, fuck it. It's just too hard to swim against the tide. I'll just turn around and float.'" Bamforth's commitment to keeping READDI patent-free had imbued Edwards with something he hadn't felt in a long time: hope that the drugs developed by Baric and the other scientists involved in the initiative would be distributed across the world rather than hoarded by the wealthy elite. Against all odds, Edwards's vision of an enterprise producing future-facing medicines for the public good was starting to materialize.

When Edwards was happy, Bamforth was happy, though he still had to worry about the bottom line. Bamforth had lost track of the number of presentations he had given pitching the initiative, though it was somewhere in the hundreds. Universities, government funders, foundations—Bamforth had pitched them all. Everyone had listened keenly, attentively, but none had yet to sign on. The goodwill and momentum were great signs, but he admitted that he had his doubts. "I can't lie," he said. "I'd be more comfortable if someone had written me a large check."

The efforts to launch READDI closely mirrored Bob Brunham's failure with SAVI, his 2003-era SARS vaccine consortium. SAVI, of course, had been undone by the original SARS virus being driven back into the wild without the need for a vaccine. Brunham had spent months and then years waiting for a second wave of SARS to arrive, which would have given his vaccine candidates a reason to be brought to market. Bamforth had been trying to sell READDI for an entire year. In that time, an armada of vaccines had arrived to save the world from SARS-CoV-2, their effectiveness so overwhelming that it was easy to assume that the pandemic was almost over. He was facing the real possibility that vaccination drives in rich countries would give funders the false sense that the millennia-old fight against coronaviruses could be easily won, despite the warning signs that SARS-CoV-2 was becoming endemic in poor countries.

There had always been a slim window to sell READDI: while the COVID-19 pandemic was waning but was not so fully over that people had lost sight that another one might arrive at any moment. Bamforth could feel the window closing. "Don't get me wrong, I think we should always continue to try to get this in place," he said, "but you can't keep pushing it forever." READDI, just like SAVI, was at risk of seeing its world-shaking vision be undone by short-term gains in the fight against the *Coronaviridae*.

By May 2021, Bamforth's exhaustion was evident. His usual low-key chumminess had faded, and he looked like he was lacking sleep. Consumed by the endless emails, calls, and Zoom pitches for funders, and egged on by Edwards to hold the line on patents, Bamforth worried that his deepest suspicions were right: that an open-science model for drug discovery simply didn't work. He believed in the project; more to the point, he believed in Ralph Baric. But at the root of Bamforth's work was also an undeniable desire to achieve something of value that could prove that, after a long and lucrative career in the pharma sector, his second act in academia could be just as consequential. Bamforth, in his typically conciliatory way, understood Edwards's concerns. Still, patents, and profit, might just be an indelible part of the fabric of scientific innovation.

Bit by bit, though, Bamforth clawed his way up the thorny bracket of public funding. Through his leadership role at the University of North Carolina, a jewel of the state, he scored an audience with North Carolina governor Roy Cooper. Over a Zoom teleconference, Bamforth gave him the pitch: READDI was the only way to protect humanity from the next pandemic and the one after that, and so on. In his pitch, Bamforth leaned heavily into Baric's track record of discovery. Baric, Bamforth reminded the governor, was the scientist who had brought remdesivir to market, creating a billion-dollar drug for Gilead. It was Baric who had also proven that EIDD-2801, Merck's little brown pill, could effectively destroy every known coronavirus. In both cases, despite that work, Baric had had no control over what the pharma companies that owned the drugs did to get them to market. Now, what if, Bamforth asked, Baric and his team had the resources to control every aspect of drug discovery, from identifying specific protein targets (the work that Edwards and the Structural Genomics Consortium excelled at), to testing drugs in animal models and multiple human cells, to finally getting it into human trials and out to the parts of the world that needed it the most? Now that, Bamforth suggested, would really put North Carolina on the map: it would be forever known as the state where humanity's best protection against pandemics had been created. Cooper, who had been following READDI since the summer of 2020, was keen to support the program. It helped that Ralph Baric was a known quantity in the state, having brought tens of millions of research dollars into the UNC system. While Cooper was sold on the vision, the governor explained to Bamforth that he needed to find the funds to do it. And as the political headwinds in the United States began to shift, he saw a way to make it happen.

On November 3, 2020, the election of Joe Biden as president rewrote the national response to COVID-19. Biden had campaigned on a promise to buoy the slumping economy with a financial recovery plan that would propel America out of the pandemic. When he took the oath of office on January 20, 2021, it signaled that American states would suddenly be awash in cash. In early March, the $1.9 trillion American Rescue Plan was signed into law. By early May, Governor Cooper had decided how to spend the $5.7 billion that the Biden administration had allocated to North Carolina, and it looked like he was going to keep his word on READDI.

Bamforth, his usual calm comportment slipping, anxiously awaited word of the deliberations. Then, days before the official announcement was made, he quietly got word: the governor had earmarked $50 million in funding directly to READDI. Bamforth, who had feared that the window was closing on the project, breathed a massive sigh of relief. Beyond powering READDI's search for broad-based antivirals, the public funding meant that he, Baric, and Edwards had leverage to ensure that they wouldn't have to rely on patenting—what Edwards described as "barter money"—to keep the project afloat. Before the pandemic, the whole idea of investing in fighting viruses that didn't yet exist, and doing so through an open-science model, would have seemed politically absurd. It had, instead, become a key part of America's recovery. It was just another example of how much the pandemic had changed things: the world had finally come to believe in Ralph Baric's vision of the future.

May 2021, Bethesda, Maryland

"I've kept thinking that over this last decade, every two years we've had a crisis," Barney Graham said, referring to the emergence of MERS, avian flu, Ebola, Chikungunya, Zika, and finally SARS-CoV-2. "It seems like somebody would start finally getting the idea." Maybe this time, he figured, they just might, citing an estimate that the U.S. economy had lost sixteen trillion dollars over the course of the pandemic's first year. "How much can you afford to spend on this work, per year, if you could save sixteen trillion dollars? I mean, spending five to ten billion dollars a year, which is probably what you would need, seems trivial." And yet, for all the success of the COVID-19 vaccines, Graham was frustrated by the typically short-sighted human response, which was to focus exclusively on the here and now at the expense of the bigger picture. He had tracked two hundred and fifty vaccine programs around the world, some of which had shown promise, but most were so far from the finish line that they were essentially useless in ending the COVID-19 epidemic. The obvious question—with an even more obvious answer—

was whether humanity really needed two hundred and fifty vaccines against one virus. The absurdity of the situation blew Graham's mind. "I would rather see those two hundred and fifty groups split into groups of ten and be working independently on each of the twenty-six virus families known to infect humans. So that twenty years from now, we will have all the information we need ahead of time, kind of like we did for coronavirus this time," he said, before offering a dire warning as he reflected back on the past year. "If this had been a different virus, we would probably still be doing antigen design right now instead of immunizing people." Humanity had gotten lucky, Graham contended: The many parts and functions of the *Coronaviridae*, including the family's ubiquitous spike protein, which Graham used to design the pinioned antigen used in most COVID-19 vaccines, had been largely mapped out before SARS-CoV-2 emerged. The virus that emerged just happened to be a member of the family that Ralph Baric had been obsessing over for forty years.

Graham had been around long enough to see multiple pandemics engulf our species. The common thread among all of them, he explained, was that they had started as regional diseases. Even HIV, which saw no signs of abating as the twenty-first century wore on, began as an obscure, regional zoonotic infection in the 1960s. "If we'd had the tools to recognize it," Graham contended, "it would have stayed this little, small problem in Central Africa." That pointed the way forward: the only way to stop pandemics was by taking care of the needs of the many. "We are so interconnected and interdependent that a problem anywhere is a problem everywhere." Not only did world powers need to commit to regional surveillance, but the deaths of people in faraway countries should be cause for alarm for all of us, even if for purely selfish reasons. This public health ethos extended to vaccinations, too. With the emergence of variants of concern across parts of the world where vaccinations lagged, it was abundantly clear that failing to vaccinate everyone across the world would leave our species vulnerable to endless transmission cycles as the SARS-CoV-2 virus mutated to escape the best weapons humanity had mustered. As Graham well knew, with only half the world immune, it would be only a matter of time before the coronavirus returned in a new and terrifying form.

May 2021, San Diego, California

Some days, Davey Smith felt like he was master of the wave. On others, that he was drowning under it. Since the aborted hydroxy-chloroquine study, ACTIV-2 had, remarkably, tested five different drugs in under a year, two of which (the immunosuppressant barici-tinib and the monoclonal antibody bamlanivimab) had been given emergency-use authorization by the FDA. It was a whirlwind of study protocols, recruitment targets—Smith had finally reached over two thousand participants—and conference calls with funders, trial sites, and his collaborators. Any one of those drug trials should have taken eighteen months to set up and at least two to three years to complete, but the COVID-19 pandemic had warped all percep-tion of time. For most people under lockdown, days stretched list-lessly on. For scientists responding to the pandemic, a frenetic pace engulfed their lives as the normal barriers to research productivity were abruptly torn down. This speed was necessary, but it had prob-ably taken years off Smith's life. And as the string of drugs registered in ACTIV-2 continued to lengthen, he began to seriously wonder if the trial would ever end.

"That is my nightmare," Smith said. "It's crazy, it's absolutely crazy, and I don't know if I would have said yes if I knew that this was going to be how it was all the time." He had found ways to adapt. He had begun working out during his phone calls and had, absurdly, begun taking two meetings at the same time. He shrugged. "I try to run one and then listen for my name in the other one and then just follow up a few minutes later." The enormity of the trial was mind-boggling, but Smith had kept his funders happy. Unlike some of the other COVID-19 trials funded by the NIH, Smith said obliquely, "We're not breaking anything," meaning that the trial hadn't hit any major problems that had derailed it. He chalked this up to the initial trauma of the hydroxychloroquine trial's failure to launch, which had spurred him to anticipate any and all threats to its success.

ACTIV-2 may have been running smoothly, but Smith saw only system failure in the broader response to COVID-19. "People are astounded that infectious diseases always hit the most vulnerable among us," he said, "and yet we never shore up our public health and our healthcare systems. We just don't realize that living in a country of haves and have-nots based on healthcare only makes us all sicker." His anger burned. "The things we failed to do?" He laughed. "I was keeping a list for a while." Evidently, it had grown too long and too depressing to maintain.

June 9, 2021, University of North Carolina, Chapel Hill

The pandemic had been pulling Baric closer and closer into the spotlight, a place from which he instinctively shied away. Despite the sudden onslaught of interest in him and his work, he was still just a basic scientist studying the minuscule and expansive universe of viruses and the cells they invaded. Throughout most of his career, when he described what he did, he could tell that most people couldn't fathom why he cared so much about the genome size of an arcane viral family. Baric had early on come to terms with the fact that his opportunities to make a difference, to translate what he did with tissue samples, mice, and chimeric viruses to help save lives in the real world, would be few and far between. The science would have to be its own reward.

Despite trusting his instincts to follow a lonely and circuitous path, the world had other ideas. Like a chance mutation leading to greater viral fitness, Baric's desire to unlock the mystery of the *Coronaviridae*'s impossibly large genomes had unforeseeable repercussions. It led him to the discovery of the genomic proofreader, shared by all coronaviruses, and from there to a plan to circumvent it. And that, ultimately, had led him to EIDD-2801, or molnupiravir, the pan-coronavirus antiviral drug that seamlessly absorbed itself into the family's genomes, blinding the proofreader and causing error catastrophes that ended the endless cycle of invasion, domination, and replication.

Baric had been obsessively testing EIDD-2801 in as many ways as possible: in mice, in multiple human respiratory cells, and against the panel of coronavirus strains he harbored in his lab, which encompassed the whole range (volumes one through fifty of the encyclopedia of the *Coronaviridae*). The drug remained undefeated. By June 2021, though, a final piece of the puzzle was ready: the results of Baric's clinical trial of EIDD-2801 had arrived.

Though his trial of the drug had enrolled only 202 COVID-infected people, the results were so dramatic that they left little doubt as to the drug's efficacy. The participants were first randomly sorted to receive the drug or a placebo. Then, crucially, to avoid the issues with remdesivir failing to work against late-stage COVID-19 disease, the groups were given either a placebo or EIDD-2801 within seven days of infection. Shockingly, at day three, only 2 percent of people who were given the drug still had virus in their system, compared to 17 percent of people given the placebo. Two days later, EIDD-2801 had so effectively eradicated the coronavirus that Baric and the other scientists running the study could detect no sign of it. It was an unmitigated victory and incontrovertible proof that a cure for the pandemic of the present and the future had finally been found.

On June 9, 2021, the Biden administration announced that it was spending $1.2 billion on almost two million doses of EIDD-2801 as a stockpile against the current coronavirus pandemic and in preparation for future ones. It was a remarkable turn of events, all the more so because Merck's Phase 2/3 large human trial of the drug hadn't yet been completed. Instead, the decision was based entirely on the sizable body of evidence that Baric and Mark Denison had quietly amassed over years.

What had started as a quixotic quest in 2013 to find promising antivirals within a sea of hundreds of thousands of candidates had culminated in a billion-dollar commitment to Baric's idea of preventing once-and-future pathogenic coronaviruses. The results of his human trial capped his four decades of obsessive and oblique experimentation, which had finally burst out into the real world in grandiose fashion.

Baric, loath to overplay his hand, was ecstatic. "That's so incredibly rewarding," he said as he let the gravity of the moment sink in. "If you have a broad-spectrum antiviral, you can get it to the site of the outbreak quickly," no matter what coronavirus might be involved, "and you can get it into the healthcare workers before they get infected and infect other people." Containing outbreaks was all about timing, and coronaviruses were no different. Typically, there was a four-day period between exposure to human coronaviruses and the onset of clinical disease, and the amount of virus in a person's system was highest before symptoms started to show. This asymptomatic gap had been the reason SARS-CoV-2 could spread so efficiently and, consequently, elude containment. But with the federal stockpile, EIDD-2801—easily transportable and orally administered—could be immediately given to everyone on the scene, either in the days immediately before or after they were infected. And that, Baric was sure, meant that any future coronavirus would have its transmission window abruptly shut. The U.S. government's decision to stockpile had been a bold strike against the next pandemic. But Baric and Denison had, with the rigor of their science and the creativity of their vision, made that decision easy.

Baric knew that the drug the world would know as molnupiravir alone would not be enough to fully protect humanity. But between the United States' stockpile announcement and READDI's publicly funded commitment, Baric had, in his humble and self-effacing way, fully imprinted his ideas onto the world. And this meant that after a long career of chasing his coronavirus dreams, he could finally face the undeniable truth: his work was going to save countless lives, now and into the future.

With the knowledge that his ancient adversary never tired, he knew the battle was far from over. Still, in the moment, even Baric recognized what he had done. "You offer people hope, right?" It had been a while.

EPILOGUE:
THE INVISIBLE SIEGE

January 28, 2021, Jade Boutique Hotel, Wuhan

MARION KOOPMANS WAS GOING STIR CRAZY. SHE HAD SPENT two full weeks holed up at the Jade Boutique Hotel, cordoned off from the outside world in a city as far from her home in Rotterdam as it could be. The room, laid out in muted sepia tones and appointed with faux Art Deco touches (transparent lamps and a curved chaise-longue), was luxe enough, but anywhere can feel like a prison if you aren't free to leave. Freedom, though, was a relative term: Koopmans was in Wuhan, the pandemic's most famous city, at the behest of the World Health Organization–China Joint Global Study of the Origins of SARS-CoV-2. As she well knew, having been on a WHO mission to Saudi Arabia to investigate MERS, these trips were hardly breezy affairs. The higher the premium placed by the country's government on reputational damage, the more intricate the planning became. As the Xi Jinping–era Chinese Communist Party came under intense scrutiny after a worldwide pandemic emerged on its soil, its anxiety about the mission was plainly telegraphed by the rigidity of the schedule and the phalanx of handlers assigned to the esteemed visitors.

Koopmans, well aware of the political machinations that dogged the mission on all sides, was steadfast. She was in Wuhan for a scientific investigation into the origins of the pandemic; the rest was noise. She and the other nine WHO scientists had haggled for months with Chinese government scientists over the scope of the mission, eventually agreeing to investigate four scenarios that could explain the origins of SARS-CoV-2. The first was that the novel coronavirus had emerged via direct zoonotic transmission from an animal reservoir, most plausibly a bat, to humans. The second was that the virus had made its way from its natural reservoir through an intermediate host and, only then, into humans; while some believed this intermediate host to be a pangolin, no one really knew. The third was the "cold chain" hypothesis, whereby the virus hadn't emerged from direct contact with wild animals at all, but through frozen animal products that were shipped into China from abroad. The fourth, and considered the most politically charged, was a scenario whereby SARS-CoV-2 was introduced into the general human population after escaping from a lab.

The scenarios neatly encapsulated both the totality of two decades of science on novel pathogenic coronaviruses and the prevailing political fault lines that had emerged along with SARS-CoV-2. Spillover from live or recently killed animals, either directly or through an intermediate host, had been the culprit for both SARS and MERS, and most scientists, Koopmans included, believed it to be the most likely source for SARS-CoV-2. Still, this presented a mystery: despite testing hundreds of thousands of animals far and wide, the closest relative to SARS-CoV-2 ever identified in bats was RaTG13, the coronavirus discovered by Shi Zhengli in China's southern Yunnan Province in 2013. SARS-CoV-2 and RaTG13 were 96 percent genetically identical, but still different enough that the bat coronavirus couldn't efficiently bind to human ACE2 receptors and invade, conquer, and commandeer cells. This made RaTG13 less an ancestor to SARS-CoV-2 and more a second cousin twice removed. There had to be another link.

The "cold chain" scenario, that the virus was imported into Wuhan in frozen packaged meat, most likely from Europe, was propagated primarily by the Chinese government and ignored by all self-respecting virologists outside China.

But the fourth scenario, the lab leak hypothesis, had taken on a life of its own. The theory suggested that SARS-CoV-2 had emerged from nature, was harvested by scientists, and then was released accidentally into a human population; in the evolving version of the theory, it was Shi Zhengli's Wuhan Institute of Virology itself that was the culprit. The lab leak hypothesis was, in its way, as simple and convenient as the cold chain hypothesis, encapsulating as it did the desire to find a scapegoat for the pain, fear, sickness, and death that the virus had caused the human race. But it had the added benefit of plausibility: one of the last-known SARS infections, in fact, was caused by a laboratory accident in August 2003 in Singapore, months after the original virus had been driven back into the wild.

There was also circumstantial evidence. The Wuhan Institute of Virology, which had been undertaking gain-of-function testing on bat coronaviruses in a similar vein to Ralph Baric's SARS-mouse chimeras, was under ten miles away from the Huanan Seafood Wholesale Market where the first outbreak of COVID-19 occurred. It wasn't much to go on, especially given that Wuhan was a capital of exotic dining and a destination for wild animals shipped from Southern China, but Koopmans and the team agreed to look into it. After reviewing the available data, the WHO scientists concluded that the viruses held in the Wuhan Institute of Virology were too dissimilar to have "caused" SARS-CoV-2, even if subjected to the most extreme gain-of-function experiments. At a final press conference before Koopmans's return home, the findings were announced: the WHO team had concluded that SARS-CoV-2 had very likely emerged from bats via an intermediate animal host and that it was extremely unlikely that it was the result of a lab leak. Seated at a long table set in front of a massive projector screen, Koopmans fielded questions for over an hour, responding to queries from journalists in brief, clipped tones. It was, overall, an anodyne affair.

When she returned home to Rotterdam, she faced a barrage of

criticism of the WHO's handling of the mission. The world had been intently watching her and the other scientists move through Wuhan, and it was predictable that most would be dissatisfied with their conclusions. Koopmans was unmoved. "What is the alternative?" she said. "If you really want to understand more about what happened, either you have a country do it all by itself, and you wait and see what comes up. Or you say, 'Hey, step aside. We will do it ourselves.'" Having China conduct the investigation on its own would inevitably have led people to assume that the country's government was hiding something. Trying to investigate the origins without Chinese government participation would likely have heightened China's instinctive mistrust of the global community. Both, she said, wouldn't have led the team any closer to the truth. The only other option was to work collaboratively with the Chinese government to try to dig up as much data as they could.

Nevertheless, media outlets took the mission participants to task for their refusing to stand up to China, for the WHO team member Peter Daszak's ties to the Wuhan Institute of Virology (he had been jointly funded to carry out bat coronavirus experiments with Shi Zhengli), for failing to announce a smoking gun, and for having taken a year to arrive at the site of the initial outbreak. Koopmans acknowledged that the mission was imperfect, but she saw no other path. "Given how these things work," she said, referring to the intense geopolitical pressures the mission was under, "I think this is probably the best we can do."

Ralph Baric disagreed. In an open letter he published with seventeen other scientists in *Science* shortly after the mission's conclusion, he lambasted the mission and its members for failing to live up to their scientific responsibilities. While the WHO team had been given access to data, all of it had been collected and summarized by Chinese scientists first; the WHO team hadn't had a chance to analyze any of it firsthand. Worse, despite the team's stated apoliticism, Baric saw clear bias in their final statement, which discounted a laboratory incident as an extremely unlikely introductory event despite a lack of clear evidence one way or the other.

The issue for Baric was that the experiments that Shi Zhengli and others had done on SARS-like bat coronaviruses and chimeric pathogenic CoVs, both of which required biosafety level 3 labs in the United States, could be done in China in much less secure BSL-2 labs. Baric was keenly aware of how important these designations were. Over the years, his BSL-3 lab had experienced ten potential exposures resulting from mouse bites, dropped vials of virus, and other accidents (though none resulted in a leak or in anyone's becoming ill). Each time an incident occurred, he reported it and tried to improve his lab's protocols. "The press takes it as a sign that you're sloppy in the laboratory," Baric said, but he knew better: human beings were human beings, and there was an error rate to everything they did. The best he could do was to avoid the same error twice.

Baric had a million questions about the WHO mission. In the United States, the minimal requirement for protection in a BSL-3 lab was a N95 mask. (Baric took no chances, equipping his staff with full Tyvek suits, pressurized hoods, thick aprons, and two sets of latex gloves to ward off mouse bites.) So, why had Shi worked with SARS-like bat coronaviruses at a BSL-2 lab when it was well known that some of these viruses could infect human cells? China's BSL-2 requirements may have been more stringent than those in the United States, and if that was the case, the level of safety went up; if it wasn't—if Chinese standards didn't minimally even require masks—then it went down. These were critical questions that should have been asked of the Chinese scientists, but to Baric's frustration, none of these details were in the mission report. "There's an inherent risk in exploring viruses that live in a zoonotic world," he cautioned, "and if you're going to work with them, you need to carefully consider all of the safety conditions." The lab, he intoned, was not a static place.

Surprisingly, Baric's misgivings about the WHO mission's lack of transparency didn't ultimately convince him of the lab leak hypothesis. "I still think it's most likely a natural occurrence," he said, "but the Chinese government needs to think about whether or not they made a good decision in refusing to share more data." Just as he had done throughout his career, Baric understood the long embrace of coronaviruses and their human hosts and the many ways (mutations and politics among them) in which pathogen and host became more and more tightly bound to each other. This long view extended forward, and Baric was convinced that the COVID-19 pandemic would fuel a rush toward synthetic viral recombination at labs across the world. The last thing he needed, as he entered the fifth decade of his work dissecting and reanimating chimeric coronaviruses, was an influx of amateurish virologists making mistakes and ruining the field for everyone.

Though the rise of SARS-CoV-2 had transformed Baric's pandemic-ending dreams into reality, one thing hadn't changed: he still wanted years, many more years, to probe and manipulate the *Coronaviridae,* organisms that acted as simple automatons and maddeningly complex pathogens all at once. And that made him more disappointed, perhaps, than anyone in the world about the basic scientific failures of the WHO-China mission's attempts to uncover how SARS-CoV-2 had first emerged. But he also took solace knowing that the work would continue as it always had. Science always found a way.

July 1996, China-Burma border, Yunnan Province

Years before the new age of pathogenic coronaviruses began, the epidemic triangle had already shifted to extreme angles. Nowhere was this more evident than in the knotted jungle landscape of Yunnan Province in Southern China, where coronavirus-infected bats roosted by the millions.

The border between the Communist superpower's southern province and its client state was more concept than reality. Across from China's southern Yunnan Province, the region on the Burmese side was controlled by the Shan people, an ethnically and culturally distinct group that had carved out their own semiautonomous state within the territory controlled by the military junta. It seemed a lopsided fight: the Burmese military forces were outfitted with weaponry (shotguns, rocket launchers, heavy artillery, and ballistic missiles) provided by China, but the combination of dense, mazelike tropical rain forest and the Shan's elite firefighting skills forced the junta back on its heels time and again, their military humiliated in retreat.

Chris Beyrer was at the border to collect blood. Then a junior HIV epidemiologist at Johns Hopkins University, he had come to the China-Burma border to investigate an explosive AIDS epidemic that had flared among people who injected drugs and who lived along heroin-trafficking routes in the region. He had heard whispers that the Shan State economy ran on heroin and that profits from their drug trafficking, which supplied not only the region but huge swathes of North America, had been financing their conflict. Unable to officially work in Burma, with the dictatorship closed to the world, Beyrer figured he'd travel the countries that bordered it to find answers. But when he arrived in the region, to his surprise, he found that the route to the epicenter of the action was freely accessible: from China's southern Yunnan Province, it cost only five dollars to enter Shan State. Once inside, Beyrer realized that heroin was just a tiny piece of the puzzle.

"Once you get into black-market economies," he said, "people do not want to use cash. Burma's banks were all sanctioned anyway, so they couldn't." As he entered Shan State, he realized that it wasn't just that the insurgency was claiming sovereignty over territory. It had also created its own independent economic system. "Heroin is a currency, opium is a currency, people are, too," Beyrer said. "Guns are traded directly for opium to support the insurgency." But it didn't stop there. "The heroin economy was related to illegal logging and wildlife trafficking, so there was a strong connectivity there." Connectivity was an understatement: the largest wildlife market in all of Southeast Asia was in Tachileik, a hectic and dusty frontier town located deep in Shan State, near the Thai border. "It had animals and animal parts from Burma," Beyrer recalled, "but also from Indian wildlife parks and Nepali wildlife parks, that were being smuggled across. And what was it all doing in Tachileik? From there, everything was being smuggled into China, the final destination market." China had a thriving traditional medicine market and exotic food scene; Beyrer had seen it firsthand. There were, he recalled, "exotic wildlife restaurants, where you can eat snake and bat and pangolin and civet and all these other animals." From Tachileik, the wildlife—traded for heroin, cash, weapons, lumber, antiquities, or women—were transported across the Shan-controlled border and into China via Yunnan Province, barreling down the same highway that led to the Tongguan Mine and the Shitou caves, where Shi Zhengli would, two decades later, find hundreds of SARS-like coronaviruses in the aftermath of the deaths of the three miners.

Beyrer wanted to know more. Traveling with a documentary filmmaker from the BBC, he agreed to conduct a covert mission to capture the illegal wildlife trade in action. Posing as a Westerner interested in buying wildlife parts, he entered the Tachileik market in search of a sale. "It was enormous," he recalled, "with stall after stall of animals, both dead and alive." In one cloistered wood-paneled room, he counted forty clouded leopard skins. One of the smallest of the big cats, the endangered clouded leopard is a canopy hunter, living exclusively in intact rain forests, where it hunts monkeys and birds amid the treetops. Forty clouded leopards, Beyrer realized, meant that a large swath of rain forest had been clear-cut.

There were cases full of hornbill heads, tiger parts, rhino horn. Beyrer was heartbroken. Amid the corpses, he eventually found one living animal. "It was alive in a cage," he said, "and I wasn't sure what it was. It was kind of a small thing; it looked sort of like a weasel or a ferret, but it was definitely alive. So, I bought it." The man who sold it to him was surprised, but he offered no objection. After haggling back and forth, Beyrer paid cash—Thai *baht*—and left with the small caged animal. He then drove across the nearby Thai border and released it in a forest. "It got out of its cage, and it looked traumatized and afraid," Beyrer said. "It wasn't its territory, so you think you're going to have a *Born Free* moment, but really, it kind of came out of the cage and looked around and, eventually, you know, made its way." When he got back to the city and was able to look it up,

Beyrer discovered that the animal was a ferret-badger, one of the animals implicated in the zoonotic events that introduced SARS and SARS-CoV-2 into the human population.

Reflecting back on the episode almost three decades later, Beyrer saw it take on a whole new meaning. A past president of the International AIDS Society and a researcher at the forefront of HIV prevention, he had been recruited into Operation Warp Speed's COVID-19 prevention efforts, specifically tasked with working to improve vaccination rates among ethnic minority communities in the United States. As the COVID-19 pandemic accelerated, he came to realize that his investigation of the HIV/AIDS pandemic in Southeast Asia thirty years earlier had, unbeknownst to him, revealed to him the origins of the next pathogenic threat the world would face. "All of these animal parts and the people selling them were mixed together in ways they would never be in nature. And this little creature—we just put it in a rented Jeep and drove it across the Thai border to Chiang Rai." In retrospect, Beyrer said, he realized that there was a good chance the ferret-badger was carrying a novel coronavirus. "When we released it, it could have turned right around and bitten one of us." Neither he, an epidemiologist, nor anyone else had even thought about it. "But of course, that risk exists. You are concentrating animals from different species and from an enormous area (birds, mammals, snakes, reptiles) and putting them all in one place." Beyrer, searching for one epidemic, had inadvertently stumbled upon the seeds of another.

What Beyrer saw in Shan State and China's Yunnan Province was a singular knot in the earth tying the virome to our bodies. It was the epidemic triangle shifted to obscene angles, the relationship among pathogen, host, and environment so perverted that the only possible outcome was the emergence of a novel pathogen. It was not, however, the only one of its kind. The word *spillover* hardly captured the situation by which multiple coronaviruses (SARS, MERS, and SARS-CoV-2) were shepherded into our species. Habitat destruction, humanity's perverse relationship to nature, the trafficking of living wild animals and dead animal parts, the hunting and eating of bush meat—these knots exist across the world, all of them leading, like the MERS-infected camels in Qatar squeezed together for slaughter, to the virome's advance on our species.

It is no accident that the first two decades of the twenty-first century have seen more novel human pathogens emerge than across the entire twentieth century. A globalized world founded on a notion of progress as perpetual expansion has inevitably placed us in more and more frequent contact with lurking pathogens. Humanity has, for centuries now, been mounting an invisible siege, invading the spaces in which the hidden virome lurks. This siege has created permanent footpaths between the viral universe and our own, the traffic between the two worlds accelerating as the cycle of contact among humans, wild animals, and livestock becomes entrenched.

The only sustainable solution to this feedback loop is the radical rethinking of humanity's relationship with the natural world. The steady encroachment of humans into the habitats of animals that harbor coronaviruses, such as bats, provides the *Coronaviridae* a direct line to our species. And then there is the wildlife trade, which alone accounts for an estimated one billion points of contact between humans and animals every year. There are the hunters shooting, killing, and trapping animals; the pens and storage facilities where creatures are held prior to their final destinations; animals packed tightly together on transport trucks or other vehicles, their exhalations rising and mingling as they're escorted to slaughter; the wholesalers and their underlings selling the animals' bodies; the butchers killing them; and the meat bought by millions, cash for flesh, taken into homes and eaten. Each step of the trade affords a jumping-off point for novel viruses. Through chance mutation, some of these viruses inevitably, eventually, find purchase among our species.

Baric, in one of his more expansive moments, reflected on how coronaviruses had so efficiently positioned themselves to emerge as humanity's greatest pathogenic threat. Ultimately, it was their status as generalists—a trait he had discovered in them—that had turned them into a threat. "Generalists can infect multiple species in the same environment," he said, "which sets up a perfect scenario for spillover into other animal hosts, other mammalian hosts and human beings." It was also, Baric explained, humanity's tendency for generalism—to roam, settle, and conquer new environments and dominate all creatures—that had sparked the most recent pandemic age. "In the twenty-first century, if you overlay boundaries of wild-life and human populations and then overlay density of human populations on top of that map, you find that it's the perfect inter-face for virus emergence."

The pandemic threat looms over our species as we burrow ever deeper into places in which we don't belong. It will remain a threat even in the unlikely event that the COVID-19 pandemic was caused by a lab leak. Taken across the sum of the countless interactions of the hidden virome and humanity, blaming a lab leak for SARS-CoV-2 was akin to blaming a fire department's poor response to wildfires driven by climate change. While the fire department should save as many lives as possible, it is not the source of the fiery and worsening disaster. The only way to truly blunt the threat of future pandemics is for humanity to retrench itself from the grim task of salting the earth. Banning the consumption and trade of wildlife—which China did on February 24, 2020—will undoubtedly help. So, too, will the closure, in 2021, of the mink fur trade in Denmark and the Netherlands. In the viral universe, these are but tiny concessions, blips along the time line. "There's advantages to being a generalist," said Baric. And though he was likely speaking of the virus itself, his remark spoke to a lesson that humanity could still learn from the *Coronaviridae:* that even the smallest mutation can radically trans-form the relationship between pathogen and host. Let us hope our adaptation is just beginning.

ACKNOWLEDGMENTS

I T'S BEEN WELL OVER A CENTURY SINCE THE AGE WHEN SOLITARY SCI-entific geniuses, working with a small coterie of assistants, made world-shattering discoveries. Nowadays, the pathway to those eureka moments is through large sprawling and specialized teams numbering in the dozens or even hundreds, and is predicated on a constant train of mentorship, resource sharing, and networked intelligence. But the narratives that we as readers crave haven't undergone a similar revolution: we still, by and large, want our heroes to be individual people rather than complex groups. This is all to say that writing about science requires a shorthand that distills the work of hundreds down to the actions of a few singular figures. The scientists profiled in this book would be the first to tell you that it takes a village to make the kinds of discoveries that have allowed our species to protect ourselves from coronaviruses. And so, while this book profiles some brilliant scientific leaders, they represent merely one feature—albeit, a very important one—of broader scientific communities that have collectively taken on the burden of helping humanity understand and conquer coronaviruses.

So, too, with the writing of this book, an inordinate amount of the final vision that was put to the page is due to the efforts of people other than myself. This book came to be because of the endless imagination and ruthless precision of my editor, Kevin Doughten. His boundless energy and restless desire for the best possible work at every stage was so gratifying and is the cause of much of what is right within these pages. My agent, Kirby Kim, pushed me to find the right ways to communicate the strange collective experience of the pandemic; Kirby, thank you for that, for your generosity of spirit, and for your tireless support. Lydia Morgan at Crown, who is brilliant, was instrumental in helping to edit and structure the book, and to reveal and excise any shaky foundations. Jenna Dolan, my copy editor, was appropriately unsparing and helped enormously in making the prose shine. Bruce Walsh at House of Anansi was a source of support and inspiration throughout this project. My thanks to my readers—Jesse Shapiro, José Lourenço, Miranda Elliott, Ron Werb, and Elly Van de Walle—who provided critical insight into what worked and didn't at various stages while putting up with, at times, some truly rough prose and dubious conjectures. I also owe a huge debt of gratitude to Maddy Curry, my intrepid transcriptionist, without whom this book would not have been written.

I am particularly indebted to the many scientists who did not appear in this book but who shared their time with me, and whose wisdom helped enormously in elevating my understanding of the issues: Aaron Christensen-Quick, Alan Bernstein, Allison McGeer, Bart Haynes, Carl Dieffenbach, Cheryl Arrowsmith, Chip Schooley, Eric Sievers, Joel Wertheim, Judith Currier, Mark Heise, Meaghan Thumath, Phil Dormitzer, Rebecca Shapiro, and Richard Gold. Thanks also to Jerica Pitts, Kelly Magnus, Shannon Devine, Isabel Ramsey, Chris Chaplin, and Toni Baric for their support.

I was the beneficiary of the work of some incredible journalists who have been covering these issues for years. In particular, Helen Branswell, Kai Kupferschmidt, and Carl Zimmer have long carried out rigorous reporting on coronaviruses that served as a foundation for my own research. I also owe thanks to the Canada Council for the Arts, which provided crucial support for this project.

And finally, thanks to Miranda and Zuleika for filling my days with joy.

NOTES

CHAPTER 1

26   **a bovine coronavirus** Marco A. Marra et al., "The Genome Sequence of the SARS-Associated Coronavirus," *Science* 300, no. 5624 (2003): 1399–404.

CHAPTER 2

33   **vertebrates, invertebrates, lichens, and mushrooms** Vincent Racaniello, "How Many Viruses on Earth?," *Virology Blog: About Viruses and Viral Disease,* 2013, https://www.virology.ws/2013/09/06/how-many-viruses-on-earth/.

33   **ground for the viruses to penetrate** Robert R. Mercer et al., "Cell Number and Distribution in Human and Rat Airways," *American Journal of Respiratory Cell and Molecular Biology* 10, no. 6 (1994): 613–24.

34   **seven-month-old Dutch infant** Sahar Abdul-Rasool and Burtram C. Fielding, "Understanding Human Coronavirus HCoV-NL63," *The Open Virology Journal* 4, no. 1 (2010): 76–84.

34   **30 percent of all respiratory infections** Krzysztof Pyrc et al., "Mosaic Structure of Human Coronavirus Nl63, One Thousand Years of Evolution," *Journal of Molecular Biology* 364, no. 5 (2006): 964–73.

35   **high occurrence of sickle-cell trait** A. C. Allison, "Co-Evolution Between Hosts and Infectious Disease Agents and Its Effects on Virulence," in R. M. Anderson and R. M. May, *Population Biology of Infectious Diseases: Report of the Dahlem Workshop on Population Biology of Infectious Disease Agents* (New York: Springer, 1982), 245–67.

35   **that coronaviruses elicit** Aakash Pandey and Daniel E. Dawson, "The Shapes of Virulence to Come," *Evolution, Medicine, and Public Health* 2019, no. 1 (2019): 3.

36   **"Will Surely Cross the Sea"** "SNEEZES FOR ALL," *The* (New York) *Evening World,* December 13, 1889, 7.

37   "NEARLY 600 FATAL CASES" "HUNDREDS DIE OF THE GRIP," *The New York Sun*, December 28, 1889, 1.

39   **key changes made the viruses less deadly** Cornelis A. M. De Haan et al., "The Group-Specific Murine Coronavirus Genes Are Not Essential, but Their Deletion, by Reverse Genetics, Is Attenuating in the Natural Host," *Virology* 296, no. 1 (2002): 177–89.

39   **98 percent identical** Leen Vijgen et al., "Complete Genomic Sequence of Human Coronavirus OC43: Molecular Clock Analysis Suggests a Relatively Recent Zoonotic Coronavirus Transmission Event," *Journal of Virology* 79, no. 3 (2005): 1595–604.

40   **to eradicate cattle fever** Vijgen et al., "Complete Genomic Sequence of Human Coronavirus OC43," 1595–604.

41   **65 percent of their total structure** Lia van der Hoek et al., "Identification of a New Human Coronavirus," *Nature Medicine* 10, no. 4 (2004): 368–73.

41   **when NL63 emerged** Krzysztof Pyrc, Ben Berkhout, and Lia van der Hoek, "Antiviral Strategies Against Human Coronaviruses," *Infectious Disorders— Drug Targets (Formerly Current Drug Targets—Infectious Disorders)* 7, no. 1 (2007): 59–66.

42   **semipermanent markets moved in** Lik Hang Tsui, "Complaining About Lived Spaces: Responses to the Urban Living Environment of Northern Song Kaifeng," *Journal of Chinese History* 2, no. 2 (2018): 335–53.

42   **silver, incense, camel, and sheep** Patricia Buckley Ebrey and Conrad Schirokauer, *Song Engagement with the Outside World*, Asia for Educators, Columbia University, 2008, http://afe.easia.columbia.edu/songdynasty -module/outside-trade.html.

43   **to more than 100 million** John D. Durand, "The Population Statistics of China, A.D. 2–1953," *Population Studies* 13, no. 3 (1960): 209–56.

43   **expanding, reaching farther and farther away** Asaf Goldschmidt, "Epidemics and Medicine During the Northern Song Dynasty: The Revival of Cold Damage Disorders (Shanghan)," *T'oung Pao* 93 (2007): 53–109.

43   **Era of Three Abundances** Shi Nai'an, *Water Margin: Outlaws of the Marsh* (North Clarendon, Vt.: Tuttle Publishing, 2011).

44   *Cold Damage and Miscellaneous Disorders* Asaf Goldschmidt and Xu Shuwei, *Medical Practice in Twelfth-Century China: A Translation of Xu Shuwei's Ninety Discussions [Cases] on Cold Damage Disorders* (New York: Springer, 2019).

46   **"no epidemics either"** Lin Yutang, *The Gay Genius: The Life and Times of Su Tungpo* (Portsmouth, N.H.: William Heinemann, 1936).

46   **many thousands of years in China** Zi-Wei Ye et al., "Zoonotic Origins of Human Coronaviruses," *International Journal of Biological Sciences* 16, no. 10 (2020): 1686–97.

46   **five hundred years later, in 1580** Katherine Markle, "Influenza Pandemics of the Twentieth Century" (master's thesis, University of Vienna, 2010).

48   **took place before 1870** Barkev S. Sanders, "The Course of Inventions," *Journal of the Patent Office Society* (1936): 666.

48   **5 percent every year** Nicholas Bloom et al., "The Geography of New Technologies," Institute for New Economic Thinking, Working Paper No. 126

(2020), https://www.ineteconomics.org/uploads/papers/WP_126-Tahoun
-et-al.pdf.

CHAPTER 3

52   **ornate structures known as syncytia** Wan Chen and Ralph S. Baric, "Molecular Anatomy of Mouse Hepatitis Virus Persistence: Coevolution of Increased Host Cell Resistance and Virus Virulence," *Journal of Virology* 70, no. 6 (1996): 3947–60.

54   **disco music** Nell Greenfieldboyce, "How a Tilt Toward Safety Stopped a Scientist's Virus Research," NPR, November 7, 2014, https://www.npr.org/sections/health-shots/2014/11/07/361219361/how-a-tilt-toward-safety-stopped-a-scientists-virus-research?t=1627132154494.

58   **"replication competent genome"** Mark R. Denison, "Coronaviruses," *RNA Biology* 8, no. 2 (2011): 270–79.

60   **he shoved it—hard** Ralph S. Baric, "Episodic Evolution Mediates Interspecies Transfer of a Murine Coronavirus," *Journal of Virology* 71, no. 3 (1997): 1946–55.

62   **paper describing OC43** J. C. Hierholzer, "Protein Composition of Coronavirus OC 43," *Virology* 48, no. 2 (1972): 516–27.

62   **"were of limited value"** Hierholzer, "Protein Composition of Coronavirus OC 43."

63   **have human orthologs** Mouse Genome Sequencing Consortium, "Initial Sequencing and Comparative Analysis of the Mouse Genome," *Nature* 420, no. 6915 (2002): 520–62.

63   **three hundred million years** Joel O. Wertheim, "A Case for the Ancient Origin of Coronaviruses," *Journal of Virology* 87, no. 12 (2013): 7039–45.

CHAPTER 4

71   **didn't always pick up the virus** Junhui Zhai et al., "Real-Time Polymerase Chain Reaction for Detecting SARS Coronavirus, Beijing, 2003," *Emerging Infectious Diseases* 10, no. 2 (2004): 311–16.

CHAPTER 5

81   **almost twice its annual revenue** Pfizer, "Financial Review: Pfizer Inc. and Subsidiary Companies: Appendix A, 2019 Financial Report," Pfizer, 2019.

89   **revenues made up only roughly 1.5 percent** Committee on the Evaluation of Vaccine Purchase Financing in the United States, *Financing Vaccines in the 21st Century: Assuring Access and Availability* (Washington, D.C.: National Academies Press, 2004).

90   **for any coronaviruses, let alone SARS** Jane Qiu, "How China's 'Bat Woman' Hunted Down Viruses from SARS to the New Coronavirus," *Scientific American* 322, no. 6 (2020): 24–32.

91   **antibodies for at least one coronavirus** Wendong Li et al., "Bats Are Natural Reservoirs of SARS-like Coronaviruses," *Science* 310, no. 5748 (2005): 676–79.

92   **every continent except Antarctica** Rachel L. Graham, Eric F. Donaldson, and Ralph S. Baric, "A Decade After SARS: Strategies for Controlling Emerging Coronaviruses," *Nature Reviews Microbiology* 11, no. 12 (2013): 836–48.

93   **10 percent of sampled bats** Susanne Pfefferle et al., "Distant Relatives of Severe Acute Respiratory Syndrome Coronavirus and Close Relatives of Human Coronavirus 229E in Bats, Ghana," *Emerging Infectious Diseases* 15, no. 9 (2009): 1377–84.

93   **thirteen different coronavirus strains** S. J. Anthony et al., "Coronaviruses in Bats from Mexico," *Journal of General Virology* 94, no. 5 (2013): 1028–38.

93   **a sampling of Natterer's bats** Tom A. August, Fiona Mathews, and Miles A. Nunn, "Alphacoronavirus Detected in Bats in the United Kingdom," *Vector-Borne and Zoonotic Diseases* 12, no. 6 (2012): 530–33.

93   **and the Indian subcontinent** Nischay Mishra et al., "A Viral Metagenomic Survey Identifies Known and Novel Mammalian Viruses in Bats from Saudi Arabia," *PLoS ONE* 14, no. 4 (2019): e0214227.

96   **who died after getting infected** T. Scobey et al., "Reverse Genetics with a Full-Length Infectious cDNA of the Middle East Respiratory Syndrome Coronavirus," *Proceedings of the National Academy of Sciences* 110, no. 40 (2013): 16157–62.

96   **first, second, and third waves** Barry Rockx et al., "Synthetic Reconstruction of Zoonotic and Early Human Severe Acute Respiratory Syndrome Coronavirus Isolates That Produce Fatal Disease in Aged Mice," *Journal of Virology* 81, no. 14 (2007): 7410–23.

97   **they were designed to attack** Damon Deming et al., "Vaccine Efficacy in Senescent Mice Challenged with Recombinant SARS-CoV Bearing Epidemic and Zoonotic Spike Variants," *PLoS Medicine* 3, no. 12 (2006): e525.

99   **The cave stank like hell** Qiu, "How China's 'Bat Woman' Hunted Down Viruses from SARS to the New Coronavirus," 24–32.

100   **77 percent identical to SARS** Xing-Yi Ge et al., "Coexistence of Multiple Coronaviruses in Several Bat Colonies in an Abandoned Mineshaft," *Virologica Sinica* 31, no. 1 (2016): 31–40.

100   **"more of a curiosity"** R. Stone, "A New Killer Virus in China?," *Science,* March 20, 2014.

101   **"SARS-like CoV from another bat"** Li Xu, "The Analysis of 6 Patients with Severe Pneumonia Caused by Unknown Viruses" (master's thesis, Kunming Medical University, China, 2013).

CHAPTER 6

107   **came back negative** Ali M. Zaki, "Isolation of a Novel Coronavirus from a Man with Pneumonia in Saudi Arabia," *New England Journal of Medicine* 367, no. 19 (2012): 1814–20.

109   **"Saudi Arabia: Human Isolate"** Ali Zaki, "Novel Coronavirus—Saudi Arabia: Human Isolate," *International Society for Infectious Diseases,* ProMED, September 12, 2012, Archive No. 20120920.1302733.

109   "the Egyptian doctor" Janice Schipper, "Egyptian Doctor Loses Job in Saudi Arabia for Submitting Letter to ProMED About New Coronavirus NCoV," FluTrackers, March 24, 2013, https://flutrackers.com/forum /forum/novel-coronavirus-ncov-mers-2012-2014/saudi-arabia-corona virus/143593-egyptian-doctor-loses-job-in-saudi-arabia-for-submitting -letter-to-promed-about-new-coronavirus-ncov.

109   responsible, therefore, for spreading MERS Tariq al Meena, "Middle East Coronavirus: No Reward for Man Behind Discovery," *Gulf News*, June 8, 2013.

113   R_0 as high as 6 Sibylle Bernard-Stoecklin et al., "Comparative Analysis of Eleven Healthcare-Associated Outbreaks of Middle East Respiratory Syndrome Coronavirus (MERS-CoV) from 2015 to 2017," *Scientific Reports* 9, no. 1 (2019), https://doi.org/10.1038/s41598-019-43586-9.

117   possible transmission routes Ziad A. Memish et al., "Middle East Respiratory Syndrome Coronavirus in Bats, Saudi Arabia," *Emerging Infectious Diseases* 19, no. 11 (2013): 1819–23.

120   "we need to put this out" Lipkin quoted in Helen Branswell, "Virus Fragment from Bat in Saudi Arabia Perfect Match for MERS Virus: Study," *The Canadian Press*, August 22, 2013.

122   injected them with Botox "Camels Banned from Saudi Beauty Contest over Botox," BBC News, January 24, 2018.

122   upward of $4 million Paul Oberjuerge, "The Camel Beauty Contest," Paul Oberjeurge (website), December 7, 2012, http://www.oberjuerge.com/http: /www.oberjuerge.com/the-camel-beauty-contest/.

123   98 percent, tested positive R. A. Perera et al., "Seroepidemiology for MERS Coronavirus Using Microneutralisation and Pseudoparticle Virus Neutralisation Assays Reveal a High Prevalence of Antibody in Dromedary Camels in Egypt, June 2013," *Eurosurveillance* 18, no. 36 (2013): 20574.

123   30 million strong B. Faye, "How Many Large Camelids in the World? A Synthetic Analysis of the World Camel Demographic Changes," *Pastoralism* 10, no. 1 (2020), https://doi.org/10.1186/s13570-020-00176-z.

123   "camels are implicated" Kai Kupferschmidt, "Bat Out of Hell? Egyptian Tomb Bat May Harbor MERS Virus," *Science*, August 22, 2013, doi: 10.1126/article.24290.

129   tested positive for MERS variants Abdulaziz N. Alagaili et al., "Middle East Respiratory Syndrome Coronavirus Infection in Dromedary Camels in Saudi Arabia," *mBio* 5, no. 2 (2014): e00884-14.

129   largest in the Middle East Indexbox, "Camel Meat Market in the Middle East Is Driven by Rising Demand in Saudi Arabia," *Global Trade*, February 21, 2020.

130   raw camel milk or meat Abdulaziz N. Alagaili et al., "Reply to 'Concerns About Misinterpretation of Recent Scientific Data Implicating Dromedary Camels in Epidemiology of Middle East Respiratory Syndrome (MERS),'" *mBio* 5, no. 4 (July 2014): e01482-14.

132   "cut the moustaches short and leave the beard" Sahih al-Bukhari, Hadith No. 5892.

133 **nine million guest workers** "New Plan to Nab Illegals Revealed," *Arab News,* April 20, 2013.

133 **"sorcery"** Virginia N. Sherry, *Bad Dreams: Exploitation and Abuse of Migrant Workers in Saudi Arabia,* Vol. 15, Human Rights Watch, 2004.

CHAPTER 7

138 **five minor mutations** Martin Linster et al., "Identification, Characterization, and Natural Selection of Mutations Driving Airborne Transmission of A/H5N1 Virus," *Cell* 157, no. 2 (2014): 329–39.

139 **"it's no longer confidential"** Eryn Brown, "Studies of Deadly H5N1 Bird Flu Mutations Test Scientific Ethics," *Los Angeles Times,* December 26, 2011.

146 **"Dear Ms. Settle"** Margaret Moore, "Re: FOI Case No. 43088 (Letter to Jocelyn Kaiser)," November 13, 2004 (p. 64), Freedom of Information Office, Department of Health and Human Services, Bethesda, Md.

147 **on average twice every week** R. D. Henkel, T. Miller, and R. S. Weyant, "Monitoring Select Agent Theft, Loss, and Release Reports in the United States, 2004–2010," *Applied Biosafety* 17 (2012): 171–80.

150 **Shitou caves in Southern China** Vineet D. Menachery et al., "A SARS-like Cluster of Circulating Bat Coronaviruses Shows Potential for Human Emergence," *Nature Medicine* 21, no. 12 (2015): 1508–13.

155 **on average cost $2.6 billion** Joseph A. Dimasi, Henry G. Grabowski, and Ronald W. Hansen, "Innovation in the Pharmaceutical Industry: New Estimates of R&D Costs," *Journal of Health Economics* 47 (2016): 20–33.

CHAPTER 8

169 **the one they were inoculated against** April M. Killikelly, Masaru Kanekiyo, and Barney S. Graham, "Pre-Fusion F Is Absent on the Surface of Formalin-Inactivated Respiratory Syncytial Virus," *Scientific Reports* 6, no. 1 (2016): 34108.

180 **"insufficient sustainable funding"** Heeyoun Cho et al., "Development of Middle East Respiratory Syndrome Coronavirus Vaccines—Advances and Challenges," *Human Vaccines and Immunotherapeutics* 14, no. 2 (2018): 304–13.

181 **actually testing two unknowns** Kizzmekia S. Corbett et al., "SARS-CoV-2 mRNA Vaccine Design Enabled by Prototype Pathogen Preparedness," *Nature* 586, no. 7830 (2020): 567–71.

CHAPTER 9

187 **four hundred thousand proteins** Mattias Mann, "The Protein Puzzle," *Max Planck Research* (2017): 55–59.

188 **0.8 percent in 2000** American Association for the Advancement of Science, "Federal R&D as a Percent of GDP," AAAS, October 2020.

CHAPTER 10

197 **"it has no relationship to SARS"** Priyanka Boghani, "A Timeline of China's Response in the First Days of COVID-19," *Frontline*, PBS, February 2, 2021, https://www.pbs.org/wgbh/frontline/article/a-timeline-of-chinas-response-in-the-first-days-of-covid-19/.

199 **without official authorization** Boghani, "A Timeline of China's Response in the First Days of COVID-19."

201 **wild animals in China** Lian Pin Koh, Yuhan Li, and Janice Ser Huay Lee, "The Value of China's Ban on Wildlife Trade and Consumption," *Nature Sustainability* 4, no. 1 (2021): 2–4.

202 **at present no reason to be concerned** Eileen Drage O'Reilly, "China Hunts Cause of Mysterious Pneumonia Outbreak in Wuhan," *Axios,* January 6, 2020, https://www.axios.com/china-pneumonia-oubtreak-wuhan-e2ef8914-6bd7-46db-814d-1609d590ee07.html.

203 **only one infected person had died** Lisa Schnirring, "China Releases Genetic Data on New Coronavirus, Now Deadly," January 11, 2020, Center for Infectious Disease and Research and Policy, University of Minnesota, Minneapolis.

215 **"currently remains unclear"** Emmie De Wit, "SARS and MERS: Recent Insights into Emerging Coronaviruses," *Nature Reviews Microbiology* 14, no. 8 (2016): 523–34.

215 **probably harming patients** Lauren J. Stockman, Richard Bellamy, and Paul Garner, "SARS: Systematic Review of Treatment Effects," *PLoS Medicine* 3, no. 9 (2006): e343.

CHAPTER 11

220 **evolved to enter human cells** Kristian G. Andersen et al., "The Proximal Origin of SARS-CoV-2," *Nature Medicine* 26, no. 4 (2020): 450–52.

226 **immunity to SARS-CoV-2 might last** Corbett, "SARS-CoV-2 mRNA Vaccine Design Enabled by Prototype Pathogen Preparedness," 567–71.

228 **a small French study** Philippe Gautret et al., "Hydroxychloroquine and Azithromycin as a Treatment of COVID-19: Results of an Open-Label Non-Randomized Clinical Trial," *International Journal of Antimicrobial Agents* 56, no. 1 (2020): 105949.

230 **including SARS, from replicating** Julien Andreani et al., "In Vitro Testing of Combined Hydroxychloroquine and Azithromycin on SARS-CoV-2 Shows Synergistic Effect," *Microbial Pathogenesis* 145 (2020): 104228.

234 **Denison had begun in 2013** John H. Beigel et al., "Remdesivir for the Treatment of COVID-19—Final Report," *New England Journal of Medicine* 383, no. 19 (2020): 1813–26.

235 **known as molnupiravir** Timothy P. Sheahan et al., "An Orally Bioavailable Broad-Spectrum Antiviral Inhibits SARS-CoV-2 in Human Airway Epithelial Cell Cultures and Multiple Coronaviruses in Mice," *Science Translational Medicine* 12, no. 541 (2020): eabb5883.

237   **Baric and Denison went all out** Sheahan et al., "An Orally Bioavailable Broad-Spectrum Antiviral Inhibits SARS-CoV-2."

CHAPTER 12

241   **$24 billion annually** Marc Bain, "Covid Fears Are Disrupting the $30 Billion Global Fur Industry," *Quartz,* November 18, 2020.

242   **afflicted with winter dysentery** H. J. Van Kruiningen et al., "A Serologic Investigation for Coronavirus and Breda Virus Antibody in Winter Dysentery of Dairy Cattle in the Northeastern United States," *Journal of Veterinary Diagnostic Investigation* 4, no. 4 (1992): 450–52.

244   **choking until they died** Martin Enserink, "Coronavirus Rips Through Dutch Mink Farms, Triggering Culls," *Science* 368, no. 6496 (2020): 1169.

245   **up from 5 percent in the previous January** James Glanz, Benedict Carey, and Hannah Beech, "Evidence Builds That an Early Mutation Made the Pandemic Harder to Stop," *The New York Times,* November 24, 2020.

246   **D-amino acid in his lab** Yixuan J. Hou et al., "SARS-CoV-2 D614g Variant Exhibits Efficient Replication Ex Vivo and Transmission in Vivo," *Science* 370, no. 6523 (2020): 1464–68.

250   **1,884 participants short of his goal** "Evaluating the Efficacy of Hydroxychloroquine and Azithromycin to Prevent Hospitalization or Death in Persons with COVID-19," ClinicalTrials.gov, April 22, 2020, NCT04358068.

254   *Games of Fortune* Christiaan Huygens, *Christiani Hugenii Libellus De Ratiociniis in Ludo Aleae: Or, the Value of All Chances in Games of Fortune; Cards, Dice, Wagers, Lotteries, &C. Mathematically Demonstrated* (London: S. Keimer, 1714).

255   **to approve COVID-19 vaccines** U.S. Food and Drug Administration, "Coronavirus (COVID-19) Update: FDA Takes Action to Help Facilitate Timely Development of Safe, Effective COVID-19 Vaccines," June 30, 2020, USFDA, Silver Spring, Md.

256   **from 16 percent to 74 percent** Robert L. Atmar et al., "Norovirus Vaccine Against Experimental Human Norwalk Virus Illness," *New England Journal of Medicine* 365, no. 23 (2011): 2178–87.

257   **"we can do this this year, in 2020"** Dan Diamond, "The Crash Landing of 'Operation Warp Speed,'" Politico, January 17, 2021.

258   **"that becomes a lot of pressure"** Abby Hamblin and Kristy Totten, "Get to Know Davey Smith, Chief of Infectious Diseases at UC San Diego," *San Diego Union-Tribune,* September 15, 2020.

CHAPTER 13

266   **"testing, testing, testing"** Clark Mindock, "'Testing, Testing, Testing': Pelosi Introduces 'Evidence Based' Coronavirus Response Just Before Trump Speech," *The Independent,* March 13, 2020.

267   **5,200 per million people** Dennis Normile, "Coronavirus Cases Have Dropped Sharply in South Korea: What's the Secret to Its Success?," *Science,* March 17, 2020.

267   **crest the one-million mark** "Totals for the U.S," The COVID Tracking
Project (*The Atlantic*), 2020, https://covidtracking.com/.

CHAPTER 14

277   **a cocktail of antiretroviral drugs** Chang-Yi Wang et al., "Effect of Anti-
CD4 Antibody UB-421 on HIV-1 Rebound After Treatment Interrup-
tion," *New England Journal of Medicine* 380, no. 16 (2019): 1535–45.

281   **Regeneron's antibody cocktail, the drug appeared** "REGN-CoV2 Independent Data Monitoring Committee Recommends Holding Enrollment
in Hospitalized Patients with High Oxygen Requirements and Continuing Enrollment in Patients with Low or No Oxygen Requirements," October 30, 2020, Regeneron Pharmaceuticals, Tarrytown, N.Y.

283   **only 45 percent effective** Erika Edwards, "This Season's Flu Shot 45 Percent
Effective, an Improvement over Last Season's Vaccine," NBC News, February 20, 2020.

283   **hundreds of thousands of lives** Stephanie Watson, "What's Herd Immunity,
and How Does It Protect Us?" WebMD, December 13, 2018.

285   **only as few as 60 percent of people** Sarah M. Bartsch et al., "Vaccine Efficacy Needed for a COVID-19 Coronavirus Vaccine to Prevent or Stop an
Epidemic as the Sole Intervention," *American Journal of Preventive Medicine*
59, no. 4 (2020): 493–503.

288   **efficacy in protecting against COVID-19** Lindsey R. Baden et al., "Efficacy
and Safety of the mRNA-1273 SARS-CoV-2 Vaccine," *New England Journal of Medicine* 384, no. 5 (2021): 403–16.

288   **between 89.3 and 96.8 percent** Baden et al., "Efficacy and Safety of the
mRNA-1273 SARS-CoV-2 Vaccine."

289   **54.8 percent and 80.6 percent efficacious** Merryn Voysey et al., "Safety and
Efficacy of the ChAdOx1 nCoV-19 Vaccine (AZD1222) against SARS-
CoV-2: An Interim Analysis of Four Randomised Controlled Trials in Bra-
zil, South Africa, and the UK," *The Lancet* 397, no. 10269 (2021): 99–111.

291   **more tightly** Andrew Rambaut et al., "Preliminary Genomic Characterisa-
tion of an Emergent SARS-CoV-2 Lineage in the UK Defined by a Novel
Set of Spike Mutations," ARTIC Network, December 9, 2020, https://viro
logical.org/t/preliminary-genomic-characterisation-of-an-emergent-sars
-cov-2-lineage-in-the-uk-defined-by-a-novel-set-of-spike-mutations/563.

292   **the site where the virus met the cell wall** Wei Bu Wang et al., "E484K Mutation in SARS-CoV-2 RBD Enhances Binding Affinity with hACE2 but
Reduces Interactions with Neutralizing Antibodies and Nanobodies: Binding Free Energy Calculation Studies," *bioRxiv* (2021), https://doi.org/10
.1101/2021.02.17.431566.

292   **60 percent more transmissible** Centers for Disease Control and Prevention,
"SARS-CoV-2 Variant Classifications and Definitions," CDC, July 6, 2021,
https://www.cdc.gov/coronavirus/2019-ncov/variants/variant-info.html.

292   **evaded treatment from bamlanivimab** Centers for Disease Control and
Prevention, "SARS-CoV-2 Variant Classifications and Definitions."

294   **conferred 60 percent protection** N. Andrews et al., "Effectiveness of

COVID-19 Vaccines Against the B.1.617.2 Variant" (2021), https://doi
.org/10.1101/2021.05.22.21257658.

CHAPTER 15

302   **25 percent less likely to die** John H. Beigel et al., "Remdesivir for the Treatment of COVID-19," *New England Journal of Medicine* 383, no. 19 (2020): 1813–26.

303   **an open-label design** Jeremy Hsu, "COVID-19: What Now for Remdesivir?" *BMJ* 371 (November 2020): m4457, https://doi.org/10.1136/bmj .m4457.

303   **"remdesivir for COVID-19"** "WHO Guideline Development Group Advises Against Use of Remdesivir for COVID-19," *BMJ* 19 (November 2020).

306   **all develop resistance to remdesivir** Maria L. Agostini, "Coronavirus Susceptibility to the Antiviral Remdesivir (Gs-5734) Is Mediated by the Viral Polymerase and the Proofreading Exoribonuclease," *mBio* 9, no. 2 (2018): e00221-18.

306   **one of his affiliations** Angela Wahl et al., "SARS-CoV-2 Infection Is Effectively Treated and Prevented by EIDD-2801," *Nature* 591, no. 7850 (2021): 451–57.

307   **remdesivir's ability to stop viral replication** Agnieszka M. Szemiel, "In Vitro Evolution of Remdesivir Resistance Reveals Genome Plasticity of Sars-Cov-2," *bioRxiv* (2021), https://doi.org/10.1101/2021.02.01.429199.

CHAPTER 16

327   **could detect no sign of it** William Fischer et al. "Molnupiravir, an Oral Antiviral Treatment for COVID-19," 2021, Cold Spring Harbor Laboratory, preprint at *medRxiv* (June 17, 2021), https://doi.org/10.1101/2021.06.17 .21258639.

EPILOGUE

331   **driven back into the wild** Kathryn Senio, "Recent Singapore SARS Case a Laboratory Accident," *The Lancet Infectious Diseases* 3, no. 11 (2003): 679.

332   **despite a lack of clear evidence** Jesse D. Bloom et al., "Investigate the Origins of COVID-19," *Science* 372, no. 6543 (2021): 694.

338   **one billion points of contact** Beatrice Jin, "How to Stop a Pandemic Before It Starts, Illustrated," Politico, n.d. 2020.

339   **on February 24, 2020** Koh, Li, and Lee, "The Value of China's Ban on Wildlife Trade and Consumption," 2–4.

INDEX

ABOUT THE AUTHOR

Dan Werb, Ph.D., is an award-winning writer and epidemiologist whose work has appeared in *The New York Times, Salon, Believer Magazine,* and many other outlets. He is an assistant professor in the Division of Infectious Diseases and Global Public Health at the University of California, San Diego, and in the School of Public Health at the University of Toronto. Werb is the author of *City of Omens: A Search for the Missing Women of the Borderlands* and has written dozens of epidemiological studies investigating the link between social conditions and the spread of disease.

Twitter: @dmwerb